(INTERMEDIATE)

压花艺术

（中级）

[美] 朱少珊　著

国际压花协会推荐参考书

中国林业出版社

图书在版编目（CIP）数据

压花艺术：中级 / 朱少珊著. -- 北京：中国林业出版社，2017.3
（2024.8重印）

ISBN 978-7-5038-7897-8

Ⅰ.①压… Ⅱ.①朱… Ⅲ.①压花－装饰美术－技法（美术）Ⅳ.①J525.1

中国版本图书馆CIP数据核字(2017)第057540号

策划编辑：何增明　印芳

责任编辑：印芳

中国林业出版社·环境园林出版分社

出　　版	中国林业出版社（100009 北京西城区刘海胡同7号）
电　　话	010－83143565
发　　行	中国林业出版社
印　　刷	河北京平诚乾印刷有限公司
版　　次	2017年5月第1版
印　　次	2024年8月第3次印刷
开　　本	710毫米×1000毫米　1/16
印　　张	9.5
字　　数	300千字
定　　价	58.00元

前言
Preface

　　爱默生曾经说"大地以花来微笑"。花，无论何种，都是美，带给人们快乐。每个爱花之人都梦想花儿永开。压花艺术给我们一种方法，不仅令花永生，还可以带来艺术的生命。无论您想要放松，获得一些乐趣，或是认真地想以压花作为您的艺术媒介或事业，这套实用的书都能有效地指导您探索压花艺术的美妙世界。对于有经验的压花艺术家，这套书将成为很好的参考资料，我想您会同意这套书很漂亮。

　　我希望我可以带领您探索更深入地用压花抒发您的感情。从这套书中学到的良好技术将能带您走得很远就像它让我走得很远一样。不管您住在哪里，也不管您的生活方式如何，这些来自大自然的美总能让您享受一些平和安静的时刻。我希望通过展示我的艺术和展示如何创作它们的细节，能让读者体会到我对生活的积极的态度。如同伊曼努尔·康德所解释的，美是"无目的的快乐"，我希望美丽的压花艺术带给您快乐。

　　非常感谢我的儿子庆耀先校正我的英语版本，还要感谢我的先生庆承侃一直在我旁边支持我。

　　我也非常感激国际压花协会的会员们和我广大的国际压花朋友们。我从大家无私的分享中学习了很多，和协会一起成长。通过研讨会和课程，网络交流，我与全世界的朋友们共同学习了，探索了，也分享了想法、方法和花材。

　　本书包含了一些材料更深入的用法，并介绍了压花艺术的艺术一面。花压得也比初级困难一点。这本书的目标是首先让大家看到每种媒介或材料都有许多使用方法。如果初级是广泛地探索各种各样的材料，中级就是试图带领您深入探索如何使用其中一种材料。它还为那些没有艺术训练的读者提供艺术概念，以了解每个作品设计背后的"为什么"。它旨在为您创作自己的压花艺术设计做准备。

Ralph Waldo Emerson once said, "[the] Earth laughs in flowers". The beauty of flowers bring smiles to all of us. Every flower lover dreams of having flowers last forever. Pressed flower art provides us a way to preserve flowers to last and also to give it artistic life. Whether if you want to relax and have some fun or you are serious in pursuing pressed flower art as your artistic medium or your career, you will find this set of practical books useful to guide you in exploring the wonderful world of pressed flower art. I think you will agree that the pictures in these books are beautiful to look at as well. For experienced pressed flower artists, this set of books will serve as good references.

I hope that I can lead you to exploring deeper in expressing your emotions with pressed flowers. The skills that you will learn from this set of books will carry you far as they have been for me. The beauty of nature is the best way of allowing you to enjoy some peace and quiet moments, no matter where you live and what your life style is. I hope to express a positive attitude towards life by presenting my art and showing the details to create them. As Immanuel Kant explained about beauty being "an object of delight apart from any interest", I hope the beautiful pressed flower art will bring you pleasure.

Many thanks to my son Marcus Ching for proofing my English version. Also thanks to my husband Douglas Ching who has always been by my side supporting.

My gratitude is also toward the World Wide Pressed Flower Guild members, and my vast pressed flower friends throughout the world. I have gained so much knowledge from everybody's sharing of ideas and information. I have grown with the guild. Through conferences, classes, and online exchanges, I have learned, explored, and shared ideas, methods, and real pressed flowers with friends worldwide.

This book contains a little more depth on some materials. It also introduces you to the artistic side of pressed flower art. The flowers pressed are a little more difficult than in Book 1. The goal for this book is to let you see that there are many ways to use a medium/material. If Book 1 is exploring the wide variety of materials, Book 2 is trying to guide you to explore in depth of how one material is used. It also provides artistic concepts for those who does not have art training to understand the "whys" behind each project design. It aims to prepare you to create your own designs artistically.

Kate Lim

2017.04

目录
Contents

四季压花艺术欣赏
Four Seasons Pressed Flower Art Appreciation — 009

压花的多面性 The Many Faces of Pressed Flower — 016
工艺美术压花 Pressed Flowers in Craft — 016
压花艺术 Pressed Flowers in Fine Art — 017

木质作品
Wooden Projects — 019

蕨类木镜框 Ferns wooden frame — 020
　木器先前准备 Preparing Wood Objects — 021
　手染蕨类 Color Enhancing Ferns — 021
　制作原木压花镜框 Making Pressed Fern Wooden Frame — 022
压花木笔筒 Pressed Flower Pen Holder — 023
压花木纸巾盒 Pressed Flower Wooden Tissue Box — 025
原木压花牌 Wooden Pressed Flower Plaque — 028

树脂胶
Resin projects — 033

时光宝石 Glass Top — 035
底托 Dish finding — 038
无框 Frameless — 041
模具 Molding — 043
镂空加金属印
Metallic Stamping with Backless Frame — 045

背景方法
Background Methods — 051

软粉彩 Soft Pastel — 052
　均色/渐变　Blending/gradation — 054
　纸张阻挡　Paper Blocking — 055

罩白 Masking	056
线状 Ray	059
直接涂抹 Direct paint	060
橡皮擦线 Erased lines	061

丙烯 Acrylic　　062

印 Stamping	063
大理石纹 Marbling	065
塑胶卡片方法 Plastic Card Method	069
纸撑 Paper roll	069

水彩 Watercolor　　071

色块 Color Mass	072
均染 Color wash	074
海绵 Sponge	075
喷雾 Spray	076
玛莎纸 Masa Paper	077
盐 Salt	078

染料 Dyes　　079

塑料袋 Easy plastic bag method.	080

其他 Others　　085

设计基本原则
Basic Design Principles　　087

设计基本元素 Basic Design Elements	088
压花艺术设计须知 Need to Know Pressed Flower Art Design	090
确定主题 Determine Theme	092
设计元素之间的关联 Relationships between Design Elements	093
小花瓶设计 A Small Vase Design	095
诠释主题 Interpretation Of The Theme	096
压花 Pressing Flowers	096
建立重心 Determine Focal Point	097
元素与元素间的关系	099
Elements and Relationships Between Elements	099
做法 Procedures	099

真空密封与装裱
Vacuum Sealing And Framing   103

真空密封材料 Vacuum Sealing Materials   104
真空密封准备工作 Vacuum Sealing Preparation   106
抽真空步骤 Vacuum Sealing Procedures   106
贴防紫外线膜 anti-UV film Application   110
装裱 Framing   112

鲜花组合
Pressed Flower Design   117

芍药近观 Peony Close-up   118

玫瑰花束 Rose Bouquet   121
压制小型玫瑰 Pressing Small Roses   122
选择搭配花 Select Companion Flowers   122
选择背景 Select Background   122
制作步骤 Procedure   122

康乃馨小花瓶静物 Still-life Carnation Vase   124
压康乃馨 Pressing Carnations   125
组合康乃馨 Assemble Carnation   126
制作画 Making the Picture   130

向日葵 Sunflower   133
压制向日葵 Pressing Sunflowers   134
制作方法 Procedure   135

郁金香小花束 Small Bouquet of Tulips   138
压制郁金香 Pressing Tulips   139
薄花处理 Extra Step for Thin Flowers   140
组合郁金香 Assemble Tulip   141
小郁金香花束设计 A Small Tulip Bouquet Design 144

蝶舞玫瑰园 Butterflies in Rose Garden   145
压花 Pressing   146
背景 Background   146
设计要点 Design Notes   147
双层设计步骤 Procedures For Layered Work   148
完成蝶舞图 Finish the Picture with Buterflies Dancing 151

  你有没有幻想过花可以永远开下去？你可曾梦想过花从春开到冬？
  利用现代干燥技术，压花保留住花的原色。它结合艺术家们的创意，呈现出一幅幅美丽独特的艺术作品。

Have you ever wondered what if flowers would last forever? Have you dreamed about having flowers bloom from spring to the winter time?
Utilizing modern drying techniques, flowers are pressed to dry to retain their natural colors. Pressed flowers combine with artists' imagination become works of art.

01

四季压花艺术欣赏

Four Seasons Pressed Flower Art Appreciation

我离开你的时候正好是春天,
当绚烂的四月,披上新的锦袄,
把活泼的春心给万物灌注遍。
　　　　　　　——莎士比亚

From you have I been absent in the spring,
When proud-pied April, dressed in all his trim,
Hath put a spirit of youth in every thing.
　　　　　　　——William Shakespeare

丰收的秋季、暴怒的冬季，
都改换了他们素来的装束，
惊愕的世界不能再凭着他们的出产辨别
出谁是谁来。

——莎士比亚

The childing autumn, angry winter, change

Their wonted liveries, and the mazed world,

By their increase, now knows not which is which.

——William Shakespeare

可是无论小鸟的歌唱，或是万紫千红、芬芳四溢的一簇簇鲜花，都不能使我诉说夏天的故事，或从烂熳的山洼把它们采撷：我也不羡慕那百合花的洁白，也不赞美玫瑰花的一片红晕；它们不过是香，是悦目的雕刻，你才是它们所要摹拟的真身。

——莎士比亚

Yet nor the lays of birds, nor the sweet smell

Of different flowers in odour and in hue,

Could make me any summer's story tell:

Or from their proud lap pluck them where they grew:

Nor did I wonder at the lily's white,

Nor praise the deep vermilion in the rose,

They were but sweet, but figures of delight:

Drawn after you, you pattern of all those.

—— William Shakespeare

当一条条冰柱檐前悬吊,
汤姆把木块向屋内搬送,
牧童狄克呵着他的指爪,
挤来的牛乳凝结了一桶,
刺骨的寒气,泥泞的路途,
大眼睛的鸱鸮夜夜高呼:
哆呵!
哆喊,哆呵!它歌唱着欢喜,
当油垢的琼转她的锅子。
当怒号的北风漫天吹响,
咳嗽打断了牧师的箴言,
鸟雀们在雪里缩住颈项,
玛利恩冻得红肿了鼻尖,
炙烤的螃蟹在锅内吱喳,
大眼睛的鸱鸮夜夜喧哗:
哆呵!
哆喊,哆呵!它歌唱着欢喜,
当油垢的琼转她的锅子。

——莎士比亚

When icicles hang by the wall
And Dick the shepherd blows his nail
And Tom bears logs into the hall
And milk comes frozen home in pail,
When blood is nipp'd and ways be foul,
Then nightly sings the staring owl,
Tu-whit;
Tu-who, a merry note,
While greasy Joan doth keel the pot.
When all aloud the wind doth blow
And coughing drowns the parson's saw
And birds sit brooding in the snow
And Marian's nose looks red and raw,
When roasted crabs hiss in the bowl,
Then nightly sings the staring owl,
Tu-whit;
Tu-who, a merry note,
While greasy Joan doth keel the pot.
——William Shakespeare

压花的多面性 The Many Faces of Pressed Flower

压花具有多面性。

There are so many sides of pressed flowers.

首先，压花如同我们在初级中所学到，不是单纯地把花随便放在压花板里那么简单。必须注意花心、花瓣、角度等，以便压出来的花具有良好的形狀。我们需要了解如何压才能较好地保持其颜色。压花现在已经从传统的只有用花扩展到包括所有植物材料。这是科学和艺术的相遇。我们采用现代化的干燥和保鲜技术压制鲜花并保留其颜色，用艺术的手法保留其优雅的形状。

First, the pressing: it is not as simple as placing the flower on the press as we have learned in book 1. One must pay attention to the flower center, the petals, the angle, and etc.... in order to have flowers coming out with good shape. One needs to understand what methodology works the best for the particular flower in order to preserve its color. Traditionally, pressed flowers meant for flowers only but it has been extended to include all botanical materials now. This is where science and art meet. We apply modern drying and preservation technology to preserve flowers and retain their colors. We use our artistic skills to retain their graceful shapes.

压好的材料成为艺术或工艺美术的媒介或元素。作品只受限于想象力。

Materials pressed dried become the medium or the elements for art or for crafts. The products are only limited by ones imagination.

工艺美术压花 Pressed Flowers in Craft

工艺美术物品的价值众所周知，压花可使它们更加美丽。我们在初级书中学到了一些，将在这本书中学习更多。

Crafting produces tangible items that are useful. Pressed flowers make them more beautiful. We have learned several in book 1. We will learn to make more in this book.

压花艺术 Pressed Flowers in Fine Art

  压花是非常好的艺术媒介。它天然，有众多颜色和各种纹理。我们可以用它来表达情感和主张。压花也可以成为设计元素。稍后我们会讨论更多有关运用压花作为艺术媒介和设计元素的内容。

Pressed flower is a great medium for fine art. It is natural, with great colors and texture. We can use it to express our feelings and ideas. Pressed flowers can also be design elements. We will be discussing more about using pressed flowers as a fine art medium or design element in this book.

深入地观察自然，然后你就会更明白了一切。

——阿尔伯特·爱因斯坦

Look deep into nature, and then you will understand everything better.

——Albert Einstein

木制品与压花的结合很和谐，因为它们都来自大自然。这一章介绍两种不同的制作压花木制品的方式，以及如何用压花刻字。

Wood products work very well together with pressed flowers because they are both from nature. In this chapter, I will cover a couple different ways to make wood crafts using pressed flowers.

02

木质作品

Wooden Projects

蕨类木镜框
Ferns Wooden Frame

采自森林中的蕨对这个原木镜框来说真是绝配。

Ferns collected from the woods are perfect for this wooden frame.

木质作品 Wooden Projects

木器先前准备 Preparing Wood Objects

原木产品首先需要密封。因为未完成的木制品是多孔的，水和空气可以渗透。为了保护木材和压上去的花，先要密封木材保证表面无孔。密封之前，先确保木器表面平滑，粗糙的地方可用细砂纸磨一下。

密封最简单的方法就是用海绵刷涂抹聚丙炳。聚丙炳气味不强烈，对室内木器很适合。涂抹完后先等两小时让它完全干燥，再用极细砂纸擦一遍，确保所有的孔都封闭。然后再刷第二层，等待两小时确保它完全干燥再进行压花设计。

Unfinished wooden products need to be sealed first. Sealing is necessary because unfinished wood is porous. Porous means that water and air can be absorbed. We need to seal the wood to make the surface non porous in order to protect the wood and also the pressed flowers we use. Make sure the wood object is smooth before sealing. If you see a rough spot, use fine and extra fine sandpaper to smooth it out.

Using a sponge applicator and satin Polycrylic protective finish to go over the wooden object is the easiest way to seal. Polycrylic does not have strong odor. It is great for doing projects indoors. After applying the finish, let it completely dry for about 2 hours. Sand the surface with extra fine sandpaper to ensure all the pores are closed. Apply a second thin coat and wait for it to completely dry for about 2 hours before working on the pressed flower decoration.

手染蕨类 Color Enhancing Ferns

绿色的蕨类对紫外线非常敏感。虽然我们会使用防紫外线的膜来保护它，我还是强烈建议用初级第三章中所提到的那种染丝绸的染料来手染，因为没有什么保护方法可以过滤100%的紫外线。

Ferns are very sensitive to UV light. Although we will use UV rated film to protect them, I still strongly recommend hand dying them with silk dye as discussed in Chapter 3 section 3.2.4 of book 1. There is no protection which can filter out 100% UV.

制作原木压花镜框 Making Pressed Fern Wooden Frame

一旦我们已经做了所有的准备工作，制作压花木框就比较容易了。在蕨的梗上放适量的胶，将蕨叶固定在框上。然后修剪边缘。

Making the pressed fern wooden frame is relatively easy once we have completed all of the preparations. Put dots of glue on strategic spots just to hold the fern in place on the frame. Trim off extra fern on edges of the frame.

剪一块比镜框外围四周都大2.5厘米的热烫膜，揭去背面保护纸并贴在框架上。这种膜不是很黏，最好每个地方都按一下贴好。在热烫定型过程中，要保证膜不移动或起皱。把熨斗调到高温，把烫衣板垫（有海绵的一边）朝向木框，覆盖好整个木框。熨斗放上去烫10～15秒，然后放在另外的地方继续烫，注意压着后别移来移去。再趁热用软布或纸巾在压花的地方向下按可以使膜更好地附着植物。

Cut a piece of drytac film about 1 inch (2.5 cm) extra on all sides of the frame. Peel off the protective backing and mount on the frame. Try to secure the film by pressing down. The film is semi adhesive. It is important that the film will not move or wrinkle during heat setting. Set a clothing iron to high. Place an iron board cover with the foam facing the frame covering the entire frame. Hold the iron on one area at a time and then do another area. Iron one section at a time for 10−15 seconds. Do not move the iron back and forth. Press downing on the fern with soft cotton or facial tissue while the fern is still hot will let the film adhere better.

剪掉四个角，然后把多余的热烫膜往镜框后面贴，和之前一样烫好。用一把锋利的小刀把镜框口上的膜按照木框的形状切除。小心不要切坏木框。美丽的压花原木镜框就制作好了。

Cut out the corners and then fold the extra drytac to the back of the frame. Heat set using the procedures as before. Use a sharp craft knife to cut the drytac along the frame opening. Be careful not to cut the wood. A beautiful pressed flower wooden frame is finished.

压花木笔筒
Pressed Flower Pen Holder

我们还可以直接把压花贴在木器上面。当然，如同之前的木器压花作品一样，首先需要密封木材。

You can attach pressed flowers to a wooden object directly. As with any wooden project, you need to seal the wood first.

压花艺术 Pressed Flower

1. 如果你想在木材上面加点颜色，丙烯是很好的选择。方法很简单，用海绵沾一些涂上去即可。

If you want to add some color to the background, acrylic is a good choice. Use a sponge to apply the paint.

2. 亚色的Mod Podge胶对于木材来说比较适合。

Matte Mod Podge is a good choice for wooden projects.

3. 整个花材需要贴在木材表面，确保每一朵小花都贴牢，等待一小时。注意不要用过多的胶。可按花材表面挤压出过量的胶一并清理掉。

Glue the entire surface of your flowers onto the project surface. Make sure every little floret is glued down. Do not use an excess amount of glue. Wait for 1 hour for the Mod Podge/glue to dry. Press down and squeeze out extra Mod Podge/glue. Clean off the excess.

4. 在花材表面刷一层薄薄的Mod Podge胶。待干燥后再刷一层，并干燥。

Brush a thin layer of Mod Podge on top of the flowers. Wait for it to dry. Brush another coat and wait for it to dry.

压花木纸巾盒
Pressed Flower Wooden Tissue Box

压花艺术 Pressed Flower

绣球对水分非常敏感，因此所有水性的胶都应该避免使用。这个作品，我们可以使用热烫膜。之前已经学习了如何在木材表面上使用热烫膜，这里我们学习怎么把绣球贴出自然的模样来。

Hydrangea flowers are very sensitive to water. Hence all water based glues are to be avoided. We can use Drytac film on this project. Since we have already learned how to apply Drytac on wooden projects, we will concentrate on making stems of hydrangeas more natural looking.

1 选一个枝梗。
Select a stem.

2

把一些绣球花贴在枝上端成不规则的圆。注意不要贴得太圆太规则，否则看起来不自然。

Arrange some hydrangea florets on top of the stem to form an irregular round shape. Note: do not try to make it too round and smooth. It does not look natural.

3

往内贴一圈绣球,覆盖大约一半外圈的绣球,但不要很平均。
贴时要部分重叠,如果只是填补空档看起来会不自然。

Arrange another round of hydrangea overlapping about half of the outer ones but do not place them in a perfect circle.
Not overlapping but just filling the space does not look natural.

4

继续重叠添加,最后用小花瓣填好中心,完成一个花球。

Do more overlapping and closing the center with smaller flowers to finish the cluster.

原木压花牌
Wooden Pressed Flower Plaque

木质作品 Wooden Projects

如同之前，木牌需要先打磨平整，边角圆滑，密封之后才能制作压花。我们在初级第9章学过如何做对角设计，这里集中学习如何制作压花字母。

Just like before, the wooden plaque has to be smoothed out and sealed before applying pressed flowers onto the surface. We have learned how to do corner design in chapter 9 beginner book. We will be concentrating on learning how to make pressed flower letters here.

1　首先，字母可以用绘图软件颠倒过来，打印出反面的字样。这里是已经翻转好的"welcome"字样。

First we can use graphics software to flip the letters so we have the negative image. Here is a welcome sign already flipped.

2　把字母描出来或者印上你自己需要的字。准备双面贴和一些大花瓣或叶子。如果用叶子，选择叶脉细的叶子。

Trace the letters or print your own letters. You also need double sided adhesive and some large petals or leaves. If using leaves, select ones without a lot of strong veins.

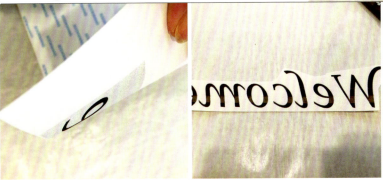

3　把双面贴贴在字母的背面（双面贴面向白色），把字母的大致形状剪下来。不要剪得太贴。

With double sided adhesive on the back side (double sided adhesive facing the blank side), cut out the general shape of the letters. Do not cut too close to the letters themselves.

4 我选择玫瑰花瓣做字母。双面贴的保护纸一点点地揭开，以防你不小心让不想粘的东西粘上去。

I have selected rose petals for my letters. Peel the double sided adhesive protection one section at a time so you can work on one section at a time to reduce errors.

5 玫瑰花瓣的颜色并不一致，注意贴的时候将你想要的色彩部分对准字母。如果担心对不准，可以先把花瓣中不要的颜色部分剪掉再贴。

Pay attention to the portion of the colors you want overlapping the letters. If not confident, you can cut off the base of the petal where color is not what you like.

6 将贴好花瓣的字母剪下来，用修剪眉毛的剪刀来剪有弧度的部分。

It is much easier to cut the curving part of the letters with eye brow scissors.

木质作品 Wooden Projects

7 排好字母在木牌上的位置，用一张纸或者一把尺子作标准，确保字母排成直线。每个字母用一小点胶固定。

With a sheet of paper or a ruler as straight line guide, position the letters on the wooden plaque. Apply a few dots of glue for each letter, then glue the letters onto the wooden plaque.

8 最后贴对角的装饰花朵。比较厚的玫瑰背后需要先修整一下再贴。背后的绿色部分尤其是翻卷成一堆的需要剪掉，这样可以减少花的厚度。

这个案例我使用了玫瑰，黄花过长沙舅，马利筋。

The back of the roses need to be flat before gluing. Trim off the green parts, especially the folded, it becomes thinner.
Rose, mecardonia, and blood flowers are used.

美是无目的的快乐。

——伊曼努尔·康德

Beauty is an object of delight apart from any interest.

——Immanuel Kant

  有谁能抗拒美丽的压花饰品？ 众所周知，每朵花都是独一无二的。 佩戴独特的从大自然来的美花，让每个人从心里微笑。

  Who can resist the beauty of pressed flower jewlery? As we all know, each flower is unique. Wearing a unique beauty from nature makes everyone smile from inside.

03

树脂胶饰品

Resin projects

在谈怎么制作之前，我想先谈论一下安全问题。市场上有很多种树脂，有些是安全的，有些则不然。不管哪种树脂，一旦完全固化，也是安全的。

Before talking about making these beautiful items, I want to discuss about safety first. There are many kinds of resin in the market. Some are safe, others pose health hazards. Resin, once completely cured, is totally safe.

美国材料测试协会（ASTM）设有树脂材料安全标准。 只有那些已经持有证书的材料，我们能够确定它是安全的。 大部分AB树脂胶在固化过程中释放有毒气体，使用时要确保在通风良好的环境中。

The ASTM International has standards for safety in materials. Only those products having received certificates are indeed safe. Most of the concerns are with two part resin which emit toxic fumes during the curing process. If you are not sure, always use caution when working with resin. Make sure you work in a well ventilated environment.

本章涵盖的所有作品都是用美国紫外线（UV）树脂完成的。 它可直接从瓶中倒出使用，不需要混合，需要紫外线才能固化。 在晴朗的日子，拿到外面晒10分钟就固化了。如果你不能保证制作的时候有大太阳，建议最好配9V或以上的紫外线灯。 一些美甲沙龙使用的灯就很好用。

All of the projects covered in this chapter are done with UV resin. UV resin can be used directly from the bottle. There is no mixing required. It requires UV rays to cure. In a nice sunny day, one can go outside and get it cured in 10 minutes. It is the best using a 9V or more UV lamp. The ones commonly found at nail salons work well for our application.

我使用的UV树脂胶与任何水质产品如丙烯或水性胶水都不兼容。 因此，注意不要与UV树脂胶同时使用上面这些产品。

The UV resin I have been using does not work with any water based products such as acrylic paint or water based glue. Therefore, make sure you do not use these products when making pressed flower jewelry with UV resin.

时光宝石
Glass Top

时光宝石很多地方都可以买到，非常容易制作。

Glass top is widely available. It is easy to get and very easy to use.

压花艺术 Pressed Flower

1

剪一块比较厚的背景纸。

Cut a piece of background paper. You need something that is heavier in weight.

滴两滴紫外线树脂胶放在背景上并涂抹开，然后将压花和叶子放在上面（为了避免空气被困在压花下，所以先放胶）。再加入2~3滴胶，确保压制的材料被胶覆盖。如果表面有气泡，可用打火机靠近表面，气泡很快就可消掉。

Put two drops of UV resin on the background and spread it. Then place the pressed flowers and leaves on the top. You must put 2 drops of resin down before placing the pressed flowers to avoid air bubbles. Add 2-3 more drops of resin making sure the pressed materials are covered with resin. Use a lighter and move a flame close to the surface to quickly get rid of air bubbles.

2

用玻璃盖住顶部，确保压花被覆盖。不要挤压，把玻璃盖在上面。如果有气泡，左右轻轻地把玻璃摇动一下消除。将其置于紫外线灯下10~15分钟即可。

3

Cover the top with the glass. Make sure the pressed flowers are all covered. Do not squeeze. Just let the glass sit on the top. Move the glass left and right a tiny bit to get rid of air bubble if any. Place it under UV light for 10 - 15 minutes.

树脂胶 Resin projects

4

把固化之后的物件取出,剪去多余的边。然后用E6000胶粘在底托上面。

Take the cured piece and trim off the edges. Use E6000 to glue the piece on the finding.

Note

许多人直接在底托上做,但是初学者大都不知道使用多少滴UV树脂胶,结果不是用得太少(内部有大气泡),就是过多(树脂在硬化过程中溢出)。用上面的方法可以确保先有一个好的设计,也是更容易和更干净的方法。

When you are not experienced, you are not sure how many drops of UV resin to use. Many resulting with too little (with large air bubble inside) or too much (with resin overflowing while being hardened). Creating the design separately first would ensure you have a good design before committing the jewelry finding. It is also an easier and cleaner way.

底托
Dish Finding

盘型的底托比较容易。

It is easier with a dish like jewelry finding.

树脂胶 Resin projects

将底托水平放置，并用指甲油染色，不要使用丙烯酸漆，等待它完全干燥。可以使用比较多的指甲油，甚至饱和。但确保底托是在水平位置，使颜色不要流到一边。

Place the finding horizontally and color with nail polish. Wait for it to dry completely. Do not use acrylic paints. It is fine to saturate the base. Just make sure the finding is level so the nail polish is not going to run to one side.

选择和底托相配的花材。

Select pressed flowers to go with the backing.

继续把底托放置在水平位置，在底托上加两滴UV树脂胶。用牙签把胶分散均匀。

Keep the finding on a level surface. Add two drops of UV resin in the finding. Use a toothpick to spread it.

压花艺术 Pressed Flower

4

加入压花。确保压花底部和UV树脂胶密接。

Add flowers in. Make sure the bottom of flowers are in total contact of the UV resin.

加入3~4滴UV树脂。使用牙签均匀地涂抹，确保压花完全浸没在树脂中。树脂应略微凸起。如果需要更多树脂，请一次添加一滴，不要溢出来。用打火机靠近表面可以很快地消掉气泡。

5

Add 3–4 drops of UV resin. Use the toothpick to spread it evenly. Make sure the flowers are totally submerged in the resin. The resin should form a slight dome shape. If you need more resin, add one drop at a time. Do not overflow. Use a lighter and move the flame close to the surface to quickly get rid of air bubbles.

6

放置在紫外线灯下固化15分钟。

Place it under UV light about 15 minutes to be cured.

无框
Frameless

这些无框树脂胶作品可以很好地展示花朵和叶子的自然形状，可以做出各种各样的饰品。

These are really nice to showcase the natural shapes of flowers and leaves. You can make a variety of items.

压花艺术 Pressed Flower

1

热裱压花，然后剪出花的形状或其他几何形状。
Hot laminate the pressed flowers, cut them out either by the shape of flower or a geometrical shape.

2

在一边加几滴紫外线树脂，使用牙签涂匀。用打火机靠近表面可以很快地消掉气泡。将其置于紫外线灯下固化。然后在背面重复第2步。
Add a few drops of UV resin on one side, use a toothpick to even it out. Use a lighter flame close to the surface to quickly get rid of air bubbles. Place it under UV light to cure. Then repeat step 2 on the back side.

3

使用E6000胶水粘吊坠环。
还可以有创意地加平底的水晶或珍珠，可使用很少量的UV胶粘，在紫外线灯下固化。
Use E6000 glue to glue on a bail.
You can be creative on adding flat back crystals or pearls. Use small amount of UV resin as glue. Cure under UV light.

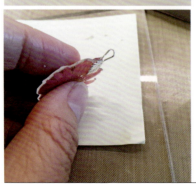

模具
Molding

网上和在实体商店有很多硅胶模具，可很容易嵌入一些压花，用树脂可创作独特的首饰挂件。

There are lots of silicone molds available both online and in stores. It is easy to embed some pressed flowers and other things in resin to create unique jewelry pieces.

压花艺术 Pressed Flower

1

往模具中加入3~4滴UV树脂。用牙签让它均匀地散开，覆盖底部。如果需要，可以添加更多滴UV树脂。用打火机靠近表面可以很快地消掉气泡。将其置于紫外线灯下固化。

Add 3-4 drops of UV resin into the mold. Use a toothpick to let it spread evenly. Add more if needed to cover the base. Use a lighter to get rid of air bubbles. Place it under the UV light to let it cure.

2

再加入2滴UV树脂，然后加入压花，确保没有空气被困在花下。然后添加更多滴的UV树脂以覆盖花材。注意分层添加UV树脂而不是一次填充。将其置于紫外线灯下固化。

如果可以，添加更多花材或其他材料，这个案例中我添加了金箔。多层创作可增添深度和更有趣的效果。

Add 2 drops of UV resin and then add pressed flowers. Make sure there is no air trapped under the flowers. Add more drops of UV resin to cover the flowers. It is wise to do the UV resin in layers rather than all at once. Place it under the UV light to let it cure.

Add more embedded items if you want. I have added gold foil to create depth and more interesting effects.

3

固化好后，把物件从模具中取出。使用指甲锉锉去粗糙的边缘，在顶部粗糙边缘处添加几滴UV树脂，把边缘所有不晶亮的地方都覆盖，置于紫外线灯下固化。

Pop the jewelry out when it is ready. Use a finger nail file to get rid of any rough edges. Add a few drops of UV resin on the top where there were rough edges. Work on the edge to ensure all dull spots are covered. Place it under UV light to cure.

镂空加金属印
Metallic Stamping with Backless Frame

这样的创意是我最喜欢的。

This method is my favorite.

压花艺术　Pressed Flower

1　取一块软陶，压平。
Take a small piece of polymer clay, roll it out.

2　盖上一块金属箔，用一张纸或纸巾快速转圈擦拭。这是为了摩擦产生一些热量。
Cover it with a piece of metallic foil. Rub it with a piece of paper or tissue in fast circular motion. This is to generate some heat.

3　提着金属箔的一个角，很快速地往对面撕，这样金属部分将留在软陶上。
Take one corner of the foil, very quickly rip it toward the opposite side. The metallic portion would be staying on the clay.

4　盖印。不需要过重但必须稳。
Stamp on the clay. Press firmly but do not exert too much pressure.

树脂胶 Resin projects

5

用包装胶带或者其他任何胶带粘去不要的金属部分。

Use packing tape or any tape to get rid of unwanted metallic foil.

6

将首饰框架放在软陶上。确保所有部位都接触好。这个步骤很重要，因为如果有任何点不接触好，树脂就会从那里漏出来。

Place the jewelry frame on the clay. Make sure all areas are on contact. This step is important since resin will leak if there is any spot not in contact.

7

滴2～3滴UV树脂胶在圈的里面并用牙签涂匀。

Put 2 −3 drops of resin and spread it with a toothpick inside the area of the ring.

8

把压花放好，确定下面没有气泡。

Place pressed flowers inside. Make sure there is no air trapped under.

9 一次加几滴UV树脂。小心不要添加太多，否则会溢出来。树脂形状会稍微凸出来，用打火机靠近表面可以很快地消掉气泡，再将其置于紫外线灯下固化。

Add UV resin a few drops at a time. Be careful of not adding too much or it will spill out. The resin will form a sight dome shape. Use a lighter to put flame close to the surface to quickly get rid of air bubbles. Place it under the UV light to let it cure.

把固化的翻过来。
重复第9步。

Flip the hardened piece over.
Repeat step 9.

树脂胶 Resin projects

11 将吊环放在坠子的前面,使用黏土支撑保持它的位置。加入几滴UV树脂,使用牙签仔细涂匀,确保树脂在吊环附近分布好。使用打火机将火焰靠近表面,以迅速消除气泡。将其置于紫外光下,让其固化。为了确保在背面的树脂也硬化,这次要照更长的时间。

Place the bail on the front of the piece. Use clay to support the bail holding it in place. Add a few drops of UV resin. Use toothpick to carefully spread. Make sure resin is all around the bail. Use a lighter to put flame close to the surface to quickly get rid of air bubbles. Place it under the UV light to let it cure. You may want to leave it longer this time since you want to make sure the resin on the back of the bail is also hardened.

  在这章将介绍一些好玩并且简单的方法制作很棒的压花艺术背景。但这不是一堂绘画课程。

  制作背景的方法太多，市面上有许多东西等待我们去发掘，这里不可能涵盖所有的方法。我在这里只有教一些简单的，希望能启发大家更深入地去探讨。

  需要说明的是背景是衬托压花的，不能喧宾夺主。刚开始不必购买很多材料，就利用手边的材料即可。

This chapter is about some fun and easy ways to create fabulous backgrounds for pressed flower art. It is not a painting lesson.

There are very many ways to create backgrounds. There is a lot available in the market for us to explore. It is not possible to cover everything here. I will only touch on a few easy ones that will inspire you to explore further.

One thing I want to mention is that the backgrounds we create should complement our pressed flower art. The background should not dominate or compete with pressed flowers.

You do not need to go buying a lot of materials. Just use what you have on hand to get started.

04 背景方法

Background Methods

软粉彩
Soft Pastel

软粉彩是压花艺术中采用得最广泛的画具，外观大多像粉笔。挑选时注意区分软、硬粉彩，因为硬粉彩也是粉笔的模样。盘式粉彩也是一种软粉彩。

Soft pastel is one of the most widely used medium in pressed flower art. It most likely comes in stick form. When buying the pastel sticks, be careful not to pick the hard paste. Instead, choose soft pastel. Pan pastel is a form of soft pastel suitable for our needs also.

最适合软粉彩用的背景材料是粉彩纸，它具备了"牙"的功能，可以咬住粉彩让背景色彩显得丰富和浓郁。不过，很多时候压花艺术需要的背景是淡彩和朦胧，所以水彩纸被广泛地运用，因为它有一定的纹理来抓住粉彩，也有硬度来支持压花。卡纸不容易抓住粉彩，但如果背景需要淡彩，还是可以运用。材料表面太光滑的不适合于粉彩。其他像无纺布也可以用来做粉彩背景。

The most common background material when working with soft pastel is pastel paper. It has "teeth" that grab the pastel powder to allow rich concentrated color. However, for most pressed flower art needs, we like to have light smoky effects to show off pressed flowers. Therefore, we will most likely use watercolor paper which has slight texture to grab the pastel

背景方法 Background Methods

color and also give stronger support for pressed flowers. Card stock does not grab pastel easily but it can be used for pressed flower art purposes since we desire lighter color backgrounds. Ultra smooth or glossy surfaces are not suitable for working with pastel. Other materials such as Japanese non-woven material can also be used to create wonderful pressed flower art backgrounds.

因为粉彩容易被碰掉，建议做完背景之后喷一下定稿剂。一些廉价的喷发胶也可以用来定稿，不过我个人使用专业的定稿剂，因为它更细更均匀。先试验一下你的喷发胶看看有没有油质，带有油质的喷发胶是不能用来定稿的。

Since pastel color can be rubbed off, I recommend you to spray fixative after you are done with the background. Some inexpensive hair spray can be used as fixative but I use professional ones that give a finer and more even coverage. Test your hair spray first to see if it contains oil. Hair spray with oil content cannot be used as fixative.

和粉彩一起使用的工具有小刀、棉花球和棉花棒。

Craft knife, cotton ball and cotton wipe are common tools used with pastels.

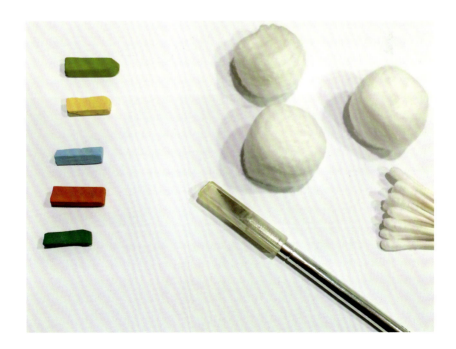

均色 / 渐变 Blending/Gradation

制作朦胧感觉背景最简单的方法就是把粉彩的粉末刮下来，然后用一个棉花球沾上粉末，涂抹在纸上。可以使用粉彩纸或水彩纸。涂抹时手法的轻重不同，色彩的深浅也不同。我们可以把不同颜色的粉末混合在一起获得各种新的颜色。把两种颜色在纸张上均匀地过渡接合也很简单，就是在两色分界的地方轻揉，直到均匀。

A simple way to create a soft smoky effect is by scraping fine powder off the pastel stick. Then use a cotton ball to dip into the pastel powder. Rub the cotton ball onto the surface of the paper. We can use either pastel or watercolor paper. By varying the force applied on the paper, we can see that we have different intensity of colors. We can mix the powders to form new colors. To blend the color boundary on paper is simple. We just gently rub the border until it blends.

背景方法 Background Methods

纸张阻挡 Paper Blocking

利用纸张制作一些特殊的效果是很有趣的玩法，我们可以采用这个方法制作很多不同的背景。

Creating this special effect using a strip of paper is a fascinating way to play. You can do many things with this technique. Let's play a little.

1 先把一张12.5厘米×17.5厘米的普通白纸任意地撕成两半。

First, we randomly tear a piece of 5×7 inch paper in two.

2 把一半的纸固定在一张12.5厘米×17.5厘米的水彩纸上。
在靠近撕痕处抹一些粉彩，然后向另外一边渐淡。
把纸取下看看，背景是不是很有趣？

Now, we place one part of the paper onto a piece of 5×7 inch watercolor paper background.
Rub some pastel colors along the tear line and gradually work the color up.
Take the paper off and take a look. Doesn't it look very interesting?

3 现在可以继续利用这个方法，为压花制作出很不错的背景。
用这个方法可以制作出很震撼的艺术作品。

Now, we can repeat the same to create a wonderful background for our pressed flower art.
You can use this same technique for some striking artwork.

罩白 Masking

● 太阳 Masking sun

从上面的练习我们看到纸张可以罩出粉彩，现在我们看看如何进一步地利用这个特点来制作这个画面的背景。

We saw how paper strips can mask out the pastel color in the last exercise, now let's take a look on how we can create a background for this piece of artwork.

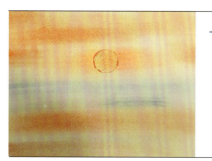

1　首先用临时胶带剪一个圆形，贴在太阳的位置。

First, we cut a piece of masking tape using a round template. Paste it onto the spot where you want the sun to be.

背景方法 Background Methods

2

然后把背景均匀上色。

Then, we color the background with pastel color and blend it well.

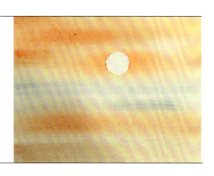

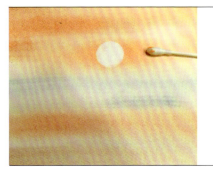

3

揭开临时胶纸，用棉花棒把太阳周围分界稍微柔和一下。

Take the masking tape off. Use a cotton tip to soften the border of the sun slightly.

● 叶子 Leaf

制作这个背景也是很好玩而且简单。

Making this background is fun and easy!

压花艺术 Pressed Flower

1. 首先找片比较有韧性的蕨，例如乌蕨，然后用手按住（从左到右，一段段来做）。

First, we find a tougher fern such as squirrel foot fern. We can just hold it in place with a finger (working one section at a time).

叶子边缘用棉花沾粉彩擦，形成叶子的形状。然后我们用棉花棒沾粉彩画出坡度。 **2.**

We pat some pastel powder with cotton along the edges of the leaf to form an outline.
Next, we use a cotton wipe with a bit of pastel color to suggest a snowy slope on the landscape.

这是又一个利用叶子和软粉彩制作的背景。

This is another example of using leaves and pastel for background.

背景方法 Background Methods

线状　Ray

有时想表现光或者风时，我们希望有线型的色彩。这其实可以在擦颜色时控制手动的方向来实现。要直线时，擦颜色时候往同一个方向即可。

Sometimes having some line of colors is what we want. It is effective in suggesting light or wind. It is all about the directions of how you move your hands when rubbing the colors. To create straight lines, you move your hand in one direction.

要弧形时，需要控制在同一个地方改变方向。

To create a swirl, you move your hand in a twisting direction and control it so you go around in about at the same direction.

直接涂抹　Direct Paint

当希望背景色彩强烈时，可以直接用粉彩棒涂抹在背景上面。也可以在背景纸上直接混色，等达到想要的色彩后再擦抹。因为一旦纸上的"牙"咬满之后，再加色就比较困难。还可以用手指直接擦抹，或者也可以让背景就带有涂抹的痕迹，也可以为你的画直接画出花瓶或其他花器。

If you desire a strong colored background, paint on the surface with pastel sticks or using the painting tool with pan pastel. You can mix colors by painting color combinations on the paper directly. Do not rub colors unless you are satisfied with the color combination because once you saturate the paper, it will be more difficult to add colors. Once you are satisfied with the color combination, you can blend colors with your fingers. Or you can just leave the background with paint strokes. You can even paint a vase or other containers for your picture.

背景方法 Background Methods

橡皮擦线　Erased Lines

橡皮可以擦出很有趣的线条或其他形状。

首先制作一个均色的背景。然后用橡皮擦出线条。用橡皮时要注意不要太轻，也不要来回擦，一次擦出线条。

Erasers can create interesting lines or other negative patterns.

First, work out a background that is blended. Next, use an eraser to make some lines. Try move the eraser confidently on the paper – not too light and not rubbing back and forth – just a single pass for the line.

丙烯
Acrylic

丙烯是一种很普及的材料，有各种质感、透明度和品质（学生级或专业级）。学生级用来做这个练习是完全可以的。

很多背景材料都可以拿来配合丙烯颜料，其中一种运用很广泛的就是卡纸。我们也可以用很多不同的手作纸和布料来试验。

Acrylic is a widely available medium that comes in several forms in terms of clarity, texture, and grades (student vs. professional). For our purpose, student grade is perfectly fine to use.

Many background materials can be used with acrylic colors. One widely used material is card stock paper. We can also experiment with many types of handmade papers as well as fabrics.

背景方法 Background Methods

印 Stamping

很多有质感的材料都可以拿来做印，我们来玩几样。

Many textured materials can be used for stamping. We will try a few.

首先取一块小气泡垫。我简单地把丙烯颜料涂在气泡这边，然后印在卡纸上。用手指仔细四周都按一遍，确定所有的气泡都和卡纸接触。把气泡垫取下，就得到了有一个有趣的背景。

First, we have a piece of small bubble wrap. I simply apply acrylic color all over the bubble wrap and then stamp on a piece of card stock. Work on the wrap surface so that all of the bubbles are on contact of the paper. Lift the bubble wrap up. Now we have an interesting stamped background.

还有很多材料可以尝试，比如可以把一张厨房铝箔团皱，然后把其中一边压在像厨房平台这样的平面上。用这个平面来印，看看是不是也很有趣？也可以尝试不同颜色混合印。

We can also try other texture surfaces. For instance, we crumple a piece of aluminum foil, and then tab it on a flat surface, such as our table, so that one side is flat. Try to use the flat surface as a stamp. Take a look how interesting it can be. We can try different colors and stamp.

能不能用植物呢？当然。让我们找一些有趣的叶子，注意最好是比较有韧性不容易流出汁液的。

How about a botanical material? Let's find a leaf with some interesting patterns. You need to look for a tougher leaf that would not bleed easily.

这里我用了一片新鲜的小玫瑰叶子，因为它比较有弹性。

I am using a mini rose leaf here. You want to work with fresh ones so it is flexible.

压花艺术 Pressed Flower

1 在叶子的背面涂一些绿色和少许金色的丙烯颜料（一定要用背面，因为叶子背面的脉才比较明显）。

I have applied some green and a little bit gold acrylic paint on the back side of the leaf (always work on the back side since the vein is more visible on the back).

2 把叶子放在要印的卡纸上面。

Place the leaf on a piece of card stock to be stamped.

3 在叶子上面放两层面巾纸，然后用手指按，注意要把叶子所有地方都按好。

Put two layers of facial tissue on top of the leaf and use your finger to press the leaf down to stamp. Make sure you have covered all areas of the leaf..

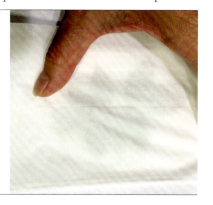

4 完成。

Finish.

背景方法 Background Methods

大理石纹　Marbling

　　我很喜欢玩这个简易大理石纹染纸。虽然它没有专业方法那么好，但是用来制作小画或卡片也绰绰有余。它几乎不需要准备，所以随时可以玩。

A very fun and easy way to make some marble paper is something I love to do. It is not as good as the professional way but it is good enough for making small pictures and cards. It requires almost no preparation so I can play whenever I have a little bit of free time.

材料	Materials
1. 白卡片纸若干、明矾1茶匙、热水4汤匙	1. A few sheets of white card stock paper、Alum, 1 teaspoon、Hot water, 4 table spoon
2. 流质丙烯颜料数种加少许水调稀	2. Fluid acrylic colors, a few colors (if your colors are thick, you need to thin it down with water)
3. 硬纸板1块、牙签数根、胶纸1条	3. One piece of card board、A few toothpicks、Strong tape, 1 piece
4. 大盆2个（最好是长方形）	4. 2 large bowls (best to use rectangular shaped)
5. 菱粉（或玉米淀粉，或太白粉）1/4杯、水4杯	5. Corn starch, 1/4 cup、Water, 4 cups

● 材料准备

把材料1的明矾用热水化开,然后用大刷子或海绵刷在纸面上。要全部覆盖，但不需要很湿。纸会卷起来，但干后会舒展开。再用重书压平。这可以前一天或前几个小时做，但不要放太久。明矾的作用是让颜料附着于纸上，隔太久，明矾的效力就会丧失。如果不用明矾，纸染出来的色彩会比较淡。

Dissolve 1 teaspoon of alum with hot water (Material Group 1). Use a large brush or sponge on one side of the card stock paper. Must have full coverage but not too wet. The paper will curl but it will expand out after dried. Use a heavy book to press paper flat. This can be done the day before or a few hours before but not too far in advance. The alum is for colors to stick on paper. It will lose its power after a while. If this step is not done, the marble paper color will be very light.

压花艺术 Pressed Flower

● 材料5准备

把材料5搅拌均匀，一边搅一边加热，直到煮沸，浆变透明。放凉。我这个案例中材料5的份量是用于可以染半张普通打印纸尺寸的盆。如果你的盆更大，需要加倍。如果买到浆衣服的浆（美国的WalMart有卖），就不需要上面的步奏。

If you can find liquid starch, (can be found in WalMart), you can skip this step. Otherwise, stir Material Group 5 together evenly and heat. You need to stir it while cooking until boil and the material becomes transparent. Let it cool. The material is for coloring 1/2 sheet of letter size paper. If you are making larger size, you need to double the amount.

● 材料3准备

裁一块硬纸板，长度小于大盆的宽。宽度大约7.5～10厘米。

把数根牙签整齐地排列在硬纸板上，一半伸出。然后用胶带粘起来，成很疏的梳子状。

Cut a piece of cardboard length to be smaller than the width of the bowl. Width is about 3-4 inches.

Line a row of toothpicks on the card board, half of the toothpick needs to be extended out. Then tape the whole thing with strong tape. It looks like very coarse comb.

1

工作台铺好废报纸，一个盆里放好清水。

Line your table with old newspaper. Fill one bowl with clean water

背景方法 Background Methods

2

另一个盆里注入浆。需要大约2.5厘米深。

Fill the other bowl with liquid starch. It needs to be about 1 inch deep.

3

刷过明矾的纸,需要裁成比盆小的尺寸。我用半张纸。
然后滴下流质的丙烯颜料。

Make sure alum treated paper is smaller than the bowl. I am using 1/2 sheet.
Then you drop some fluid acrylic colors.

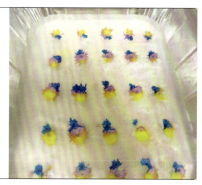

4

用粘好牙签的板子慢慢转大圈把颜色转出美丽的漩涡。

Use the toothpick tool to slowly move in circular motion to make some swirls.

5

把卡片纸慢慢放下,刷明矾一边朝下。注意中间不能有气泡。确定四周都在浆上了之后,快速提起一角,把纸放在装清水的盆里慢慢摇,把浆冲洗掉。然后晾起来,可以用晾衣服的夹子夹好,晾干。

With the alum treated side facing down, slowly place the card stock paper down to the bowl. Make sure there are no air bubbles in between. After making sure all sides are on contact with the starch, lift one corner and take it out quickly. Place the colored paper in the bowl with clean water. Shake it slowly to wash out the starch. Hang it on a clothing rack or line to dry.

压花艺术 Pressed Flower

6 干后用重书压好过夜。
每次染一张。因为颜料比较容易沉底，第二次需要再次加颜料。一盆浆可以染很多张纸，我一共染了8张，每张色彩和花纹都不一样。

Put it under heavy book overnight to flat out.
You can only marble one sheet at a time. The used paint will sink with this method. You will need to add paint for the next sheet. The starch can support many sheets per batch. I have colored 8 – 10 sheets per batch. Every sheet is different.

美丽的手染纸就做好了。

Your beautiful handmade marble paper is ready for use.

背景方法 Background Methods

塑胶卡片方法　Plastic Card Method

把丙烯颜料放在纸上端成线，然后用一张塑胶卡片往下刮。可以加多种颜色形成不同的感觉，少许金属色彩也很棒。

Spread a line of colors on one side, use a plastic card to move the colors to the other end of the paper. You can add a second color and third color for some interesting effects. A little metallic color works great in this case.

纸捲　Paper roll

把软纸巾拧成一条(也可以用布料)，沾丙烯颜料，然后像擀面一样在纸上来回转。

Twist a sheet of soft paper towel into a roll (you can use fabric also). Dip into acrylic color and roll on background paper.

2

也可以加第二种颜色。
You can add a second color also.

水彩
Watercolor

水彩是制作压花艺术背景极好的材质。

水彩有块状和膏状的。

许多不同的纸都适合做它的背景。

Watercolor is an excellent medium for making pressed flower art backgrounds.

Watercolor comes in cake form and in tubs.

Many different types of paper can be used as backgrounds.

色块 Color Mass

这是一个能够凸显一些色彩强烈花卉的有趣背景。

This is a very fun effect that can set off some strong flowers beautifully.

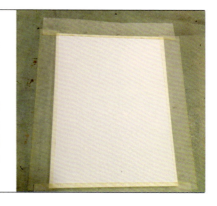

1　首先我们需要裁一张比需要的背景大约长宽都多1.3厘米的水彩纸，用临时胶带把四周大约0.6厘米贴在一个硬板上。

First, we need to cut a piece of watercolor paper 1/2 inch larger both in width and length than the size you need. Use masking tape to tape the paper on a hard board about 1/4 inch on all sides. This is to prevent the paper buckling during our process.

背景方法 Background Methods

2

● 用一支中型的笔在想要有色块处刷一些清水。然后再在没干的地方点上浓色。
● 要制作流血样子，先在色块边上擦些清水，然后慢慢地把清水和湿的色块连接。

Use a medium brush and apply clear water to the area you want the color mass to go. Then, you apply strong color by tapping the color on the wet area.
To create a bleeding effect, apply water to the edge of the color mass and slowly connect the wet area and the wet color.

3

注意，板子需要倾斜，这样水才能朝你需要的方向流。
要达成这个效果，需要在湿的地方上色。

Notice that the board needs to be slightly tilted so water will flow to the direction you want.
For this effect, we always want to apply paint on wet paper.

4

用清水和转动板子来引导颜色的运行。
当表面颜色干燥时，把背景压在厚书下面压平。

Guide the color movement by tilting the board and moving the water.
As soon as the surface color is dried, place the paper under a heavy book to flatten.

压花艺术 Pressed Flower

均染　Color Wash

　　准备足够水彩，然后用一支大毛笔，从左到右。

　　适合的背景材料很多，比如水彩纸、手造纸、桑麻纸、蕾丝等等……

　　这个案例中，我把一张宣纸团了一下，然后打开，均色染出一个很有趣的效果，干后烫平。

Prepare enough watercolors, then use a large brush, to color across the surface.

Many background materials are good to work with. Try watercolor paper, handmade paper, mulberry paper, lace, etc…

In this sample, I have crumpled a piece of rice paper. Color washed for a very interesting effect. Iron after dried.

背景方法 Background Methods

海绵 Sponge

这又是一个制作有趣背景的方法，快速而简单。

把天然海绵打湿（人造海绵孔比较密且排列整齐）。

把海绵沾一种颜色，然后随意印在纸上，洗干净海绵再换一种颜色。这个方法也适用于比较稀的丙烯颜料，或可以混合使用不同的颜料。

注意：海绵沾水彩不要太多，控制在比较干的状态；不要过分染色，留白会令画面看起来更加有趣。

This is a fast and easy way to create some interesting backgrounds.

Wet a natural sponge (manmade ones are too regular in patterns that are too predictable).

Dip the sponge in one color and stamp on the paper randomly.

Wash the sponge and then change colors.

You can also use acrylic paint with this method or mix the mediums.

Notice that you want to load the sponge lightly and keep it on the dry side.

Do not over do – leave some white spaces to make the pattern more interesting.

喷雾 Spray

用一个有喷头的瓶子可以很容易地制作各种背景。背景的均匀程度要看喷头和操作经验。流质的丙烯、墨水和流质的染料都可以拿来像水彩一样使用。使用完毕之后，记得要把瓶子，特别是喷头彻底用清水洗干净，以防喷头被堵。

很多背景材质都可以用。手造纸、桑麻纸、落水纸、薄和纸，甚至布料都是好的选择。

It is easy to create a variety of backgrounds with a spray bottle. Evenness of the spray depends on the spray nozzle and also how you control the spray bottle.

Fluid acrylic, ink, and liquid dye can be used in a spray bottle just like watercolor. Wash the bottle with clear water after use. Make sure the nozzle is cleaned to prevent clogs.

Many different background materials can be used. Handmade paper, mulberry paper, lace paper, sanwa paper, even fabric are good choices.

1 先灌一点水（不能全满）加水彩，摇晃到水彩完全融化。量视染色面积大小而定。

Fill the bottle with a little water (not all the way full), add water color, and shake until dissolved. Amount depends on the coverage area. Adjust amount accordingly.

喷头离纸面15厘米左右来喷。喷时一次按到底，而不是一点点，这样会喷出比较细腻均匀的喷雾。换颜色要先清洗瓶子和喷头，背景喷完待干再使用。

Hold the bottle about 6 inches from the paper and spray. The spray bottle will work better when you press all the way through, not just half way, for even and smooth mist. Wash the bottle when changing color. Wait until dried.

2

背景方法 Background Methods

玛莎纸　Masa Paper

玛莎纸比较坚韧，表面光滑、细腻，背面有机理。

Masa paper is a tough paper that is shiny on one side and rougher on the other side.

我制作这个材料的背景是这样的：
先在光滑的那边一角标上"光滑"。然后把纸团成一团，再打开。我团几次，让纸上有密布的皱痕。

The way we make this background is like this:
We mark one corner of the smooth side of the paper "smooth". I crumple the paper and then open; crumple it again. I do it several times so you get very fine wrinkles all over the paper.

然后把纸团放进一盆水里浸湿。把它拿出来，小心地打开。湿的纸比较容易破，因此要小心。光滑的一边朝下，放在一张厨房用纸巾上。

Then, I wet the paper by dropping the crumpled ball into a bowl of water. Take it out and open it gently since the paper is weak when it is wet. Place it smooth side down on top of a paper towel.

均色染几种颜色。你可以看到颜色会在皱褶地方比较深。

干后用熨斗烫平，我用的是光滑的一面。

Do a color wash of several colors. You will see that the colors are more concentrated on the wrinkled lines.
Wait until it is dried. Iron it. We will use the smooth side.

盐 Salt

粗盐比精盐更适合这个方法。
先预备好盐粒，在水彩纸上均匀地上色。

Sea salt with larger grains works better in this method than fine table salt. Set aside some grains of salt. Do a color wash on a piece of watercolor paper.

把盐随意地放在纸面上，等待干燥。等干后把盐粒取下。

Place the salt on the surface randomly. Wait until it is dried completely to remove the salt. It will form a snowy effect.

染料
Dyes

纺织染料和水彩相似,但更流畅。它用在各种面料包括丝绸都很好。它对薄和低、吸水纸最适合。

Fabric dyes work similarly as watercolor but is more fluid. It works great on all kinds of fabric including silk. It also works great on thin and absorbent papers.

压花艺术 Pressed Flower

塑料袋　Easy Plastic Bag Method

1

把薄和纸叠一下，放在塑胶袋里。浸湿之后把水倒掉。放几滴染料进去，用手指在袋子外面把颜料弄均匀，再戴上手套把纸拿出来，晾干。可以用吹风机加速干燥时间。

Fold a piece of sanwa paper and put it into a sandwich bag. Wet paper and drain water. Drop a few drops of dye into the bag and close the bag. Use your finger to work the dye colors evenly from outside of the bag. Wear gloves to take the paper out and hang dry. You can use a hair dryer to speed the drying.

2

干后用低温烫平。
Iron with cool iron once dry to flaten out any wrinkles.

3

团一块春雨落水纸，放在塑胶袋子里，加水浸湿，倒出多余水。

You can also do a random pattern.
Crumple a piece of rain drop lace paper. Put it in a plastic bag. Wet it. Drain extra water.

背景方法 Background Methods

4
滴几滴液体染料。
Drop a few drops of colors randomly.

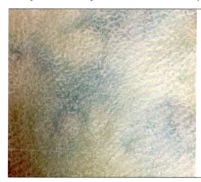

5
用吹风机吹干。纸在湿的时候非常脆弱，要小心不要把纸捅破，不要在湿的时候勉强打开。
待干后再用低温把纸烫平。
Use hair dryer to blow dry. The paper is very fragile when wet. Be careful not to poke holes. Do not force open the paper when wet.
Iron with cool iron after dried.

● 冷冻纸 Freezer Paper

1
把薄和纸烫在一张冷冻纸上面。
　冷冻纸有一层塑胶膜，并且给予薄的材料一定的支撑。
Iron the sanwa paper onto a piece of freezer paper. The freezer paper has a plastic lining and also gives support to the thin material.

2
你可以匀色染，或者用一个天然海绵沾染料染自由色块。
You can do a color wash to have even color. Or you can sponge colors on for a random pattern.

压花艺术 Pressed Flower

● 硬板 Hardboard Support

这是我最喜欢的方法。我把一张纸放在硬板上面，把整张纸打湿。然后用天然海绵沾染料拍在纸上，把染好的纸晾干。

This is my favorite way to dye. I place a piece of paper on the hardboard. Wet the paper completely.

I then sponge dye colors onto paper. Hang paper on a clothing line to dry.

用像豌豆大小的染料。我通常用饮料吸管，你也可以用小勺取。

加入清水。水的份量决定颜色的深浅。我通常用1/4量杯，大约50毫升。记得，湿的时候颜色看起来深一点。如果不确定，可以用白纸先试验一下。

Use a pea size of dye. I usually use a straw to pick up the dye but you can use a spoon too.

Mix with water. The amount of water determines the intensity of the color. I used 1/4 cup. Remember, dye color appears deeper when wet. If you are not sure, try the color out on a white paper and let dry.

把纸放在硬板上。也可以使用一块有机玻璃或任何防水的硬板。

Place the paper onto a hard board. You can use a piece of acrylic, or any hard surface that is waterproof.

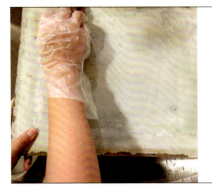

用蘸了水的海绵轻轻拍在纸上，把纸湿透。小心不要擦，否则纸在湿的情况下很容易破。

Tap a wet sponge over the paper, wetting the entire sheet. Be careful not to rub the paper since it is very weak when wet.

背景方法 Background Methods

用海绵吸一些颜色，挤在纸上，让颜色自由流下来。如果想深色一些，把多一点颜色用海绵蘸在纸上；如果想浅色一些，可以用干净海绵用清水洗掉一些颜色，不过不要直接开水龙头洗，水会把纸冲出一个洞。 **4**

Dip the sponge into a dye, squeeze the color onto paper and let it run freely. Tapping the dye over an area that you had previously colored if you desire deeper color. Lighten up dye by tapping a clean wet sponge onto an area. You can squeeze water to wash off some of the color. However, do not use running water to wash. You might create a hole in the paper from the running water.

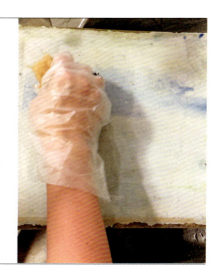

5

换颜色时，先用清水洗干净海绵，再蘸第二种颜色。可以把第二种颜色加在白色部位，也可以加在其他颜色上，就好像水彩一样。

To change colors, wash the sponge with clear water and then dip into another dye. You can add color to a white area or on top of the other color just like how you would work with watercolors.

当完成染色时，小心地取两个上端的角把纸从板上揭下来，晾干。

When you are done with colors, lift the paper carefully with the top two corners. Hang it to dry.

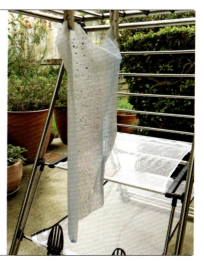

压花艺术 Pressed Flower

其他
Others

压花背景的创作无止境。运用你的想象力，很多事情都可能。以下是我尝试过的一个例子：

首先把白乳胶调稀，把一层面巾纸（通常面巾纸有两层）盖在一张紫色卡纸上，然后用毛笔沾胶水，从纸巾上面一点点地粘，便使其产生一些很有趣的褶皱。等全部干后，再把软粉彩棒横过来，把纸巾的褶皱上色，效果很奇特。

Sky is the limit for backgrounds. Use your imagination. Many things are possible. Here is an example:

I thinned down white glue and used facial tissues on a purple card. This can create wrinkles on the card. Wait until it is completely dried. Then, I run a soft pastel stick, broad side, on the tissue to highlight it. The result is very interesting.

艺术的目的是洗去我们的灵魂在日常生活中的尘埃。

——毕加索

The purpose of art is washing the dust of daily life off our souls.

——Pablo Picasso

05 设计基本原则

Basic Design Principles

设计基本元素
Basic Design Elements

　　一个美术画面，有以下基本的元素：形状，线条，颜色，空间和质地。

　　●形状：每样物体都有形状。这里形状是立体的概念。在平面设计上，通过透视达到立体的效果。一般花材是有机形体。

　　●线条：连接两点，有曲或直，宽度和方向。

　　●颜色：有三个属性。色相（颜色），纯度（颜色的鲜明度）和明度（颜色的深浅度）。

　　●空间：空间包括整个画面的空间，背景空间、中景空间和前景空间，也包括花材之间的空间与物体包含的空间。

　　质地：每样花材都有其独特的质感。我们使用的背景材料、辅助材料也有其不同的质感。

There are the following basic elements in an artwork:

Form, line, color, space, and texture

Form: every object has form. Form is a 3-D concept. In 2-D design, it is achieved by perspective. Usually plant materials have organic forms.

Line: connecting two points. Line can be straight or curved. Lines or curves have width and direction.

设计基本原则 Basic Design Principles

Color: has three attributes. Hue (the color), intensity (vividness of the color), and volume (light and darkness of the color).

Space: including background, middle ground, and foreground of the entire picture space. It also includes spaces between objects and spaces within an object.

Texture: Each plant material has its own unique texture. Background materials we use, supporting materials also have their own different texture.

举例来说，大家可以很清楚地看到下面这幅画的背景质感是比较皱的轻纱样子。在整个画面的空间内，大家也可以看到花与花之间的空间，花与叶之间的空间，以及它也暗示是立体的空间。当然我们可以看到花和叶的形状，也可以看到背景彩虹的形状。同样是兰花，因为有了层次的设计，背景的兰花和前景的兰花就有着不同的色调。因此也产生了立体的感觉。竹叶的线条，背景彩虹的线条，那也是浅而易见。至于色彩，那就更不需要说，是绿和紫以及过渡颜色。

For example, we can clearly see that the background of the picture is a wrinkled silk texture. Of the space in the entire picture, you can see the space between flowers, space between flowers and leaves, and it also implies that the space is three-dimensional. Of course, we can see the shape of flowers and leaves. You can also see the shape of a rainbow background. The orchids in the background and orchids on the front have different intensity and volume because of the design. Therefore it produces a three-dimensional look. As for the lines, the rainbow background and bamboo lines are easily understood. As for hues, there are greens and purples and many transitional colors in between.

压花艺术设计须知
Need to Know Pressed Flower Art Design

压花艺术，它是压花与艺术的结合体，它和其他平面艺术有些相似，又有些不同。

Pressed flower art includes pressed flowers and artistic relationships. Pressed flower art is somewhat similar to other 2-D arts but it is also a little different.

其他平面艺术像油画、水彩、粉彩、丙烯等，创作人从无到有，在设计好的构图中，画出各种我们前一篇所说的各种元素。

In other arts such as oil paintings, watercolors, pastels, acrylic, etc., we create from scratch. We paint all the artistic elements as we have discussed in the previous section after finishing a design of good composition.

而压花艺术包含前面所说艺术中的各种元素，像质地、形状、色调、线条、颜色等等，但这些大自然已经通过花材给予我们了，不过，我们无法把花材不受限制地放大或缩小，也无法改变其属性。我们能做的就是不断地完善压花的技术，获得不错的花材。

In pressed flower art, the flowers we have pressed already consist the artistic elements. Elements such as texture, shape, hue, line, and color have already given by Mother Nature. We only have to constantly improve our own pressing skills to obtain good floral materials. However, we cannot enlarge or shrink materials nor can we change their properties.

当然，许多人把植物剪得很细，然后只利用花材的色彩，贴出一些和其他平面艺术相似的画面来。可是细想一下，我们为什么要压花？如果只是剪剪贴贴，那我们为什么不干脆使用纸张？它也是木材或其他植物纤维造出来，而且还不会褪色。我们压花，那是源于我们对大自然的爱。我们希望把植物的美永远留下来。它的美不仅仅在于其色彩。因此，剪贴并不是不好，而是在整幅作品中，我们应该保留一些花材原来的样子，让大家能够领略到压花艺术独特的美。

Of course, many people cut or tear plant materials into tiny pieces and only use floral colors, making it similar to paints in other fine arts. However, come to think of it, why do we use pressed flowers? If we just cut and collage, then why not just use paper? Paper is also made of plant fibers, plus it will not fade. We press, it is because of our love of Nature. We want to preserve the beauty of it forever. Its beauty exists more than color alone. Thus, cutting them to tiny pieces is not bad but we should also be able to see some of the plant materials in their original form in order to appreciate the unique beauty of pressed flower art.

压花艺术的独特，在于它的天然。很多平面艺术都画植物。但是却无法画出真正植物的色彩和质感（尤其是质感）。而且，世界上的花，也没有两朵是一模一样的。发现花材的美，并巧妙地运用花材，是压花艺术的另一个独特性。巧妙地运用花材，能让我们创作出各种美丽的静物、风景、人物、动物，还能够让观者从画作中看到植物原来的样子。这等大家压花有了经验之后，会逐渐体会到。我们将在之后的章节和高级书籍中做更深的尝试。

The uniqueness of pressed flower art is because it comes from Nature. Many fine artists paint plants. However they cannot really paint the exact colors and textures of plants (especially its texture). Moreover, there are no two flowers which are exactly alike. Discovery, and the cleverly use of floral materials, is another unique factor of pressed flower art. Clever use of floral materials allows us to create a variety of beautiful still-life, landscapes, people, animals, and plants. It also allows the viewer to see the original look for the plant from the pictures. We will gradually attempt more complex artwork after we gain more experience with pressed flower art in later chapters and the next books.

确定主题
Determine Theme

现在我们了解了压花的艺术元素，那么下一步怎么开始呢？这是我的学生们开始时问得最多的问题。即便他们看见我制作的教学案例，还是觉得无从着手。

Now that we have some pressed flower art elements, what is our first step to begin? This is what my students ask the most when they first start. Although I have made samples but they still don't know how to proceed.

首先，我们需要明确主题。就像写文章一样，只有主题突出，大家读了之后才知道谈的是什么。如果只是东一朵美丽的花，西一朵同样美丽的花，会让观者无法适从，不知道这画要展现的到底是什么。

First, we need to have a clear theme. Like in writing, readers can follow a clear theme. If one just scatters a beautiful flower here and there, viewers will not be able to follow what you want to present.

建议开始设计时制定比较小型和具象的主题，等积累了经验之后，再逐渐扩大和尝试抽象的主题。

I recommend starting with smaller and figurative themes. Expand the topic and try abstract themes after one accumulates more experience.

设计元素之间的关联
Relationships between Design Elements

了解了压花的艺术元素,能够确定主题之后,我们再来看这些元素之间的关系。前人总结了一些元素之间的基本法则,让大家运用起来能够比较容易地创作出有美感和令人愉悦的画面。当然,艺术不是科学,这些法则也是让人参考的。这些法则有:统一/和谐、平衡、层次、规模/比例、优势/强调、类似/对比。

After understanding the basic elements and determining the theme, let's look at the relationships between these elements. There are some basic rules regarding design elements. Following these rules allow us to easily create a sense of beauty and visual appealing pictures. Of course, art is not science. The rules are for reference only.

These rules are: Unity/Harmony, Balance, Hierarchy, Scale/proportion, Dominance/emphasis, Similarity/Contrast.

当开始学习压花艺术设计,我们需要想这些法则。等有了经验之后,就可以不再细细考量自己的设计是否用了基本法则,是否用得好?这好比学习一种文字,先学习其语法以及句子的构造,才能写出文章来。等我们掌握了这种文字之后,就可以灵活运用了。

先了解这些法则的涵义。
- 统一/和谐。统一是整体艺术品的概念,即所有画面上面的元素都必须团结在一起成为一个艺术品;和谐是由使用相似的元素来实现，它让一件复杂的艺术品，拥有一个简单的外观。
- 平衡。平衡不等于左边一样花材，右边也用同样的花材。就好像天平上，左边放一大块东西，右边可以放许多小的来平衡一样的道理。
- 层次。层次的意思是排列元素以暗示其重要性。
- 规模/比例。规模是指物体的大小；比例是指元素之间的相对大小。
- 优势/强调。优势的意思是艺术家让一个元素脱颖而出，使它首先抓住观众的眼球而起画面的主导地位。该元素被强调。
- 类似/对比。规划一致和类似的设计很重要，这样使得焦点可见。太多的类似性是枯燥的，但没有类似，重要元素就不存在，无对比的画面是平淡无奇的。所以关键是要找到类似和对比之间的平衡。

When we start learning pressed flower art design, we should think about these principles. We don't consciously think about them as much after we gain more experience. This is similar to learning a language, we need to learn the grammar and sentence structures in order to write beautiful pieces. Once we have mastered the language, we can then be more creative.

I would like to explain the meaning of the design elements.

Unity/Harmony. Unity is the whole concept of art. It means that all the elements must unite together into a work of art. Harmony is by the use of similar elements to achieve harmony and make a complex work of art have a simple appearance.

Balance. Balance does not literally mean that you need the same flower on the left and right side of the picture. If you place a large piece on the left, you can put many smaller elements on the right.

Hierarchy. Hierarchy refers to the arrangement or presentation of elements in a way that implies importance.

Scale/proportion. Scale means the size of the object. Proportion means the relative size between objects.

Dominance/emphasis. The artist gives an element dominance by making it to stand out and drawing the viewer's eye there first. The element is being emphasized.

Similarity/Contrast. Planning a consistent and similar design is important to make the focal point visible.Too much similarity is boring but without similarity, important elements will not exist. An image without contrast is uneventful.The key is to find the balance between similarity and contrast.

小花瓶设计
A Small Vase Design

让我们运用基础美术原则来制作一幅小画，主题是夏日插花。

Let's apply the basic design principles on a small picture.The title is Summer Arrangement.

压花艺术 Pressed Flower

诠释主题　Interpretation of the Theme

这个夏日插花小品是一个很容易理解的作品。对于夏季的主题，画面选用暖色系的花朵为主花。次花可以选择强烈的对比色或同一颜色主调。对于这个主题，我设计了一个向日葵插花代表夏天。

This small summer arrangement is easy to understand. To imply the summer theme, the picture uses warm color flowers as the main flower. Complementary flowers can either have strong contrast in color or in the same color theme. For this theme, I have designed a sunflower arrangement to represent summer.

压花　Pressing Flowers

真的向日葵用在一个小图上会太大。蟛蜞菊（南美蟛蜞菊）是一种很常见的园林植物。它是向日葵家族的一员。如果我们需要长的很像向日葵的小花，这是一个很好的选择。

Real sunflowers are too large for a small picture. Wedelia or trailing daisy (sphagneticola trilobata) is a very common landscape plant. It is in the sunflower family. It is similar to a sunflowers form or shape.

1

2

采集刚刚开的花如图右边的花朵那样。当花瓣开始卷，如同左边，那就是开过一阵子了。

Collect flowers that just bloomed like the one on the right. When petals start to curl like on the left, then it has been blooming for a while. We should avoid using old blooms.

3

压正面花时把梗剪掉，注意不要剪太多伤了花材。

Cut the stem off and press face down for full face front view. Make sure not to cut too much and damage the flower.

设计基本原则 Basic Design Principles

4 侧面花需要刨半，然后压两个侧面。
For side views, slice the flower in half and press each half.

5 所有梗在花剪掉后刨半。
Slice all stems in half (once removed of flowers).

材料	Materials
1. 压好的蟛蜞菊、叶子、铁线蕨、六倍利 2. 12.7cm×17.8cm　水彩纸 3. 软粉彩（土黄）	1. Pressed wedelia, leaf, maidenhair fern, lobelia 2. 5×7 inch (or 12.7 × 17.8 cm) watercolor paper 3. Soft pastel (warm yellow)

建立重心　Determine Focal Point

　　重心往往是在画面的中心，因为我们的眼睛往往都是先看中间。然而，它不应该在正中间，令画面呆板。我们可以试图在这幅画上标上线。绿线是画面长度和宽度的一半，绿点为画面的正中心；蓝色线是在画面长度和宽度的1/4。橙色线在画面长度和宽度的1/8。通常情况下，重心是在中间1/8线，我已经标有透明蓝色椭圆。椭圆区域内的任何点（但避免绿色中心点）都可以成为画面的重心。不是所有的主花都是平等的。只有一个是焦点。这个焦点花不应该有任何其他花阻挡。它通常是前置的。其他花卉要么在这朵花后面，要么在侧面。我把我画面的重心标明红色。之所以不把焦点设在图片的中心是因为这样会给人一种非常稳定的感觉，这对于引导观众的眼睛读您的画作起反效果。焦点不只是让人眼睛选择性地看看。在艺术视线上，我们不希望观众的眼睛只是选择、寻找要看的。相反，艺术家的责任是引导观众的眼睛，在画面内协调地读画。

Focal point is often at the center of the picture since our eyes usually look at the center first. However, it should not be at the exact center. We can form grids for this picture. The green lines are at 1/2 of the length and width. The green dot in the picture is the exact center of the picture. The blue lines are at 1/4 of the length and width. The orange lines are at 1/8 of the length and width. Usually, the focal point is at the middle 1/8 lines. I have marked with blue oval. Use any point inside of the oval area but avoid the green dot (exact center). Not all main flowers are equal. Only one should be the focal point. This focal flower should not be obstructed by any others. It is usually front facing. Other flowers are either presented behind this main one or on the side. I have marked my focal point with red. The reason for not putting the focal point at the very center of the picture is because that gives people a very still feeling which would be counteractive in leading the viewer's eye traveling within your picture. The focal point isn't just wherever the eye chooses to see. In art, we don't want the eye to just choose to see what it wants to look for. On the contrary, it's the artist's responsibility to direct the eye, to orchestrate its movement within the picture.

设计基本原则 Basic Design Principles

元素与元素间的关系
Elements and Relationships between Elements

我们可以看到压花代表了像线条、形状、颜色和质地这些元素，而背景纸则很显然地代表了空间。

We can see that the elements such as line, form, color, and texture are mostly represented by the pressed materials. Space is obviously being represented by the background paper.

而且我是有顺序排列花材。有些是重叠在其他花的上面，即便很少，也可形成层次结构。主花—向日葵（蟛蜞菊）的色泽鲜艳，体形较大成为主导。平衡是通过不等边三角形来实现的。

We can also see that I have arranged the flowers in order. Some are overlapping others, forming the hierarchy structure. The main flower – sunflower (wedelia) is dominating with its bright color and larger shape. Balance is achieved by balancing the scalene triangle arrangement of the flowers.

不等边三角形通常用于定位主花。请看标志在例图的红线。要注意的一点是，没有一条线和作品的边缘平行。

Scalene triangles are often used to position main flowers. Take a look of the red lines marked in the example picture. The thing to pay attention to is that none of the lines should be in parallel to the edge of the picture.

做法　Procedures

我找到一片很漂亮的叶子，这是很好的花瓶材料。把两头剪一下，然后把角都修圆。

I found a beautiful leaf. It is perfect for making a vase. I cut the two ends and then round all the corners.

压花艺术　Pressed Flower

2 刮一些粉彩末到废纸上，用棉花球沾了擦在背景花瓶所要站立的地方。

Scrape some powder from a soft pastel stick on scratch paper. Use a cotton ball to dip the color and rub on the watercolor paper where the vase will sit.

3 先在瓶口铺陈几片蕨叶。

First arrange a few fern leaves around the opening of the vase.

然后插花。先确定哪一个是面对的那朵，其他两朵需要指向略微不同的方向。三角形位置应该不等边。

Then arrange flowers. Determine which one is the front facing one first, the other two needs to point to slightly different directions. The triangle position should be scalene.

4

5 现在插入最长的枝来决定设计的高度。

Now arrange the longest stem to determine the height of the design.

设计基本原则 Basic Design Principles

6 插入其他长枝来平衡设计。
Arrange the other tall flowers to balance the design.

7 最后插入一些小的陪衬花（六倍利）来获取一些有趣的对比色。签名写上日期就完成了。
Finally insert small companion flowers (lobelia) to provide some interesting contrast. Date and sign the picture to finish.

Note

完成的图放在一个文件夹中并放入一个拥有大量还原好的干燥剂的密封盒子里至少一夜，使其完全干透。这在高湿地区尤为重要。

The finished picture is placed in a folder and put into a box with active desiccant overnight to be completely dry. This is particularly important in high humidity areas.

卓越是一个持续的过程,而不是一个意外。

——阿卜杜尔·卡拉姆

Excellence is a continuous process and not an accident.

——A. P. J. Abdul Kalam

压花艺术有其科学的一面。保色就是一个非常复杂的过程,它需要阻止水分、空气和紫外线破坏我们的作品使其保持很长时间。在本章中将讨论真空密封的每一个步骤。

Pressed Flower Art has its scientific side. The preservation of pressed flowers is a very detailed process that blocks moisture, air, and UV rays from our finished works. Therefore, our works are preserved and will last long. In this chapter, I will go over every step of the vacuum sealing process.

06

真空密封与装裱

Vacuum Sealing And Framing

真空密封材料　Vacuum Sealing Materials

材料	Materials
1. 镜框纸卡开口比作品长和宽都小1.27cm。卡边宽度最少应该有2.5cm。具体尺寸视镜框大小来配	1. Mat opening should be 0.5 inch smaller both in length and width than the picture. Mat width needs to be minimum 1 inch. Actual size should be coordinated with the frame
2. 铝箔尺寸与纸卡尺寸相同	2. Mylar size is the same as the mat size
3. 花胶或中性玻璃胶	3. Resin or silicone silicone II glue
4. 薄棉垫尺寸与画面尺寸相同	4. Low pile cotton batting size is the same as picture size
5. 薄干燥板尺寸是1/4到1/2画面尺寸	5. Desiccant paper is about 1/4 to 1/2 of the size of the picture
6. 脱氧剂大约每100cm^{2}分1包（100g）	6. Oxygen absorber, about 1 (100 g) per 16 square inches
7. 厚装裱用卡纸板与双面贴	7. Mat board and double sided adhesive
8. 抽真空小泵	8. Small vacuum pump
9. 小木（或塑胶）勺	9. Small wooden or plastic spoon
10. 防紫外线玻璃或有机玻璃	10. Anti-UV Glass or acrylic
11. 饮品吸管适合小泵管子（先套好）	11. Drinking straw that fits to the vacuum pump outlet (connect them now)

真空密封与装裱 Vacuum Sealing And Framing

Note

如果没有防紫外线玻璃或有机玻璃，这里用普通玻璃，然后参看玻璃贴膜那一节。

关键是要有合适的复合型铝箔和胶以保持真空。如在我的《压花艺术》（初级）中解释的，复合型铝箔是为了阻断水气、氧气和其他气体（约100微米）厚。复合型铝箔需要有弹性，这样它可以在抽真空时，服帖地随作品后面之形状。好的抽真空复合型铝箔也防刺穿。胶是至关重要的，要确保使用的胶可以把复合性铝箔粘到玻璃或有机玻璃上。

If you don't have anti-UV glass or acrylic, use regular acrylic or glass and then read the section on anti-UV film.

It is crucial to have the correct Mylar and glue in order to hold the vacuum. As explained in my first book, Mylar should be 4mil (about 100 micron) thick in order to provide a barrier for water vapor, oxygen and other gases. Mylar should be flexible so that it can mold to the shape of the items. A good Mylar for vacuum sealing should also resist punctures. The glue is critical that it will bond glass or acrylic to Mylar.

真空密封准备工作 Vacuum Sealing Preparation

1

把双面贴贴在厚装裱用卡纸板上。

Apply double sided adhesive on mat board.

2

把带有双面贴的厚装裱用卡纸板裁成8毫米细条。

还原干燥板。

Cut the mat board with double sided adhesive into 1/4 inch (8mm) strips.

Re-activate desiccant board.

抽真空步骤　Vacuum Sealing Procedures

1/2 of the mat width.
卡纸框边一半宽度

Mat width
卡纸框边宽

1

把贴纸的保护层撕开，把硬纸板条在复合铝箔亮面，4个边缘离开边约半个卡纸框宽度的地方贴好。

Peel off the protection paper and adhere the hard cardboard strip to all 4 sides of the shiny side of the Mylar; leaving the edges about 1/2 of the width of the mat.

如果您必须连接纸板条，一定要确保没有空隙，也没有互相重叠。

If you must make a connection, make sure there is no gap or overlapping.

2

用木勺或塑料勺压紧纸板条。

Use the wooden (or plastic) spoon to burnish.

3

把还原好的干燥板和脱氧剂放在铝箔上面。

Place the activated desiccant board and fresh oxygen absorber on the Mylar.

真空密封与装裱 Vacuum Sealing And Framing

4

盖上一张棉垫。

Cover with a piece of cotton padding.

5

把画放在棉垫上。

Place the picture on top of the cotton padding.

6

把胶挤在硬纸板印子上。胶需要有连贯性。如果哪里比较细，需要补胶。

Squeeze glue on top of the card board strips in a continuous line. Make sure it is even. Add more if there are areas that are too thin.

7

把纸框放在玻璃上面，比好确保画的四边位置都对，再把玻璃放在胶上。

Place the mat on top of the glass/acrylic. Make sure you have the picture all 4 sides positioned correctly before gluing it down.

8

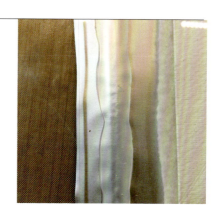

小心地两边按，把胶均匀地粘好。小心地把手伸到底下来匀胶，不能移动或把画抬起来，以防止画滑动位置。必须在硬纸板痕迹的两边都布上胶，胶不足的地方可以补上。不要着急，把胶匀好、贴好至关重要。

Press to spread the glue evenly. Do not lift the picture more than just fitting your fingers under in order to ensure that the picture would not move at this stage. Make sure the glue is spread on both sides of the card board strip mark. Add more glue to spots where not enough. Do not rush. It is very important that the glue is spread out and glued well.

9

小心地掀开一个角，把吸管（连接小泵的皮管）放入。然后把胶贴好，气泡挤出。

Carefully lift a corner to insert the drinking straw that's connected to the pump outlet. Work out all the air bubbles in the glue.

10

用手扶好饮管所在的角落，确保胶一直贴好。开泵以从画作中把空气抽出。当它达到真空状态，你可以听到泵的声音有变化。也可以从前面的玻璃上看到，所有花材、底纸等都被挤压在玻璃上。

Hold on the corner where the straw is, making sure the glue is not being lifted. Turn on the pump to draw air out from the picture. You can hear the pump's sound changing as the picture is fully vacuumed. You can also see from the front that everything is being pushed against glass/acrylic.

真空密封与装裱 Vacuum Sealing And Framing

按住饮管的角落，将作品翻过来检查。如果有气泡，用软布把气泡往吸管那个角落推来去除。

Hold the corner and turn the picture over to check. If there is any air being trapped, use a soft cloth to push the air pocket toward the corner where the straw is to get rid of it.

扶好角落，用大拇指压玻璃，其他手指在下面顶着吸管。另外一只手把吸管慢慢抽出。这样可以防止漏气。

Hold the corner, as shown above, with your thumb pressing the glass/acrylic and the rest of the fingers pressing the straw on the back. Use the other hand to slowly take the straw out. This will ensure no air leaks at this point.

确保胶仍然粘贴牢固。把作品翻过来，用木勺再次压硬纸板条各处一次。把作品放在一边过夜等待胶固化。

Make sure that the glue is still spread out evenly. Turn the picture over and use the wooden spoon to burnish the card board strips one more time. Put the picture aside overnight for the glue to cure.

鸣谢好友杉野宣雄无私的分享他的发明，并授权我教授他的方法。

Credit goes to Nubuo Sugino. I am very grateful that he has shared his invention with me and granted me permission to teach this method.

贴防紫外线膜 Anti-UV film Application

如果你没有防紫外线玻璃或有机玻璃，你则需要在外层玻璃上贴透明防紫外线膜，不然真空密封的作品仍然会受到紫外线破坏。

清洗玻璃是贴防紫外线膜最重要的工作。玻璃必须没粘绒毛和灰尘颗粒，因为贴了膜后那些微小的东西会非常明显。你可以按照防紫外线膜附带的说明来贴，也可以雇佣专业汽车玻璃贴膜公司帮忙贴。

贴好的膜通常需要2~3天才干，等干了再装裱。

If you don't have anti-UV glass/acrylic, you will need to apply clear anti-UV film to the glass for the outer layer. Otherwise the vacuum sealed artwork is still vulnerable to UV damage.

The majority of the work for applying the anti-UV film is cleaning the glass. The glass must be free of lint and of any dust particles. Those tiny particles will be very visible when film is applied. Follow the instructions that comes with the anti-UV film to apply. You can also have the film professionally applied by anyone who does car window films.

The anti-UV film usually cure in minimum 2 -3 days. You need to wait for it to cure before framing.

● 自己贴防紫外线膜需要材料　DIY Anti-UV film Materials

除了玻璃和透明防紫外线膜之外，你还需要以下材料。

Besides of the glass and clear anti UV film, you need the following materials.

材料	Materials
1. 强生无泪婴儿洗发精	1. no-tears baby shampoo
2. 瓶装水（非自来水）	2. Bottled water (not tap water)
3. 带喷头的瓶子	3. Spray bottle
4. 刮板	4. Squeegee
5. 单边刀片	5. Single-edged razor blade
6. 无毛软布或咖啡滤纸	6. Soft, lint-free cloth or paper coffee filters
7. 大毛巾用来保护桌面	7. Large towel to protect table
8. 似信用卡的塑胶卡片	8. Plastic card similar to credit card

● 自己贴防紫外线膜步骤　DIY Anti-UV film Procedure

1. 切割比框架玻璃稍大的透明抗紫外线膜（每边大约多出1.5~2.5厘米）。

2. 将婴儿洗发水1/4茶匙放入喷雾瓶中。

3. 将瓶装水倒入喷雾瓶中，摇动混合。

4. 用此混合液清洁玻璃（不要使用含有化学品的常规窗户清洁溶液），使用单刃刀片刮去任何顽固的污垢，并用无毛布或咖啡过滤器清洁玻璃。如果有任何污点需要清洁，重复。这一步非常重要。

5. 将大毛巾放在桌子上，然后将清洁后的玻璃放在上面。

6. 抗UV膜具有保护膜，它比防紫外膜更软、更薄。从一个角开始，然后分离一边。用洗发水和水的混合液把膜和保护层中间喷湿。

7. 用洗发水和水混合液彻底喷湿玻璃。

真空密封与装裱 Vacuum Sealing And Framing

8. 把防紫外线膜展开的一边贴在玻璃的一侧。

9. 一边滚动拉出保护膜，一边喷水，并将膜贴在玻璃上。

10. 如果发现有任何灰尘斑点，这时拉起薄膜，使用一块普通胶带将其粘出。

11. 使用刮板从中间开始，向4个边刮以去除水。

12. 用软布包裹塑料卡，然后在薄膜上刮，将水分去除更彻底。

13. 等待2～3天干燥。

1. Cut the clear anti-UV film slightly larger than the framing glass (about 1/2 to 1 inch (1.5 – 2.5cm) larger on each sides)

2. Add 1/4 tea spoon of baby shampoo into the spray bottle.

3. Pour bottle water into spray bottle. Shake to mix.

4. Clean the glass with this mixture (do not use window cleaning solution that contains chemicals). Use the single-edged razor blade to scrape off any stubborn dirt. Use the lint-free cloth or coffee filter to clean glass. Repeat if there is any persistent spot needing cleaning. This step is extremely important.

5. Place the large towel on table, and then place the cleaned glass on the top.

6. The anti-UV film has a protecting film. It is softer and thinner than the anti-UV film. Work one corner to separate the film. Work out to separate one side. Wet the film with shampoo and water mixture.

7. Wet the glass with the shampoo and water mixture thoroughly.

8. Line the film on one side of the glass.

9. Roll out the protecting film and spray water at the same time as you apply the film on glass.

10. If there is any lint or dust spot, lift the film up, use a piece of tape to remove it at this time.

11. Use a squeegee to start from the middle and work toward 4 sides to get rid of water.

12. Wrap the plastic card with soft cloth and then work on the film to get rid of more water.

13. Set aside for 2-3 days waiting for dry.

Note

如果你没有防紫外线玻璃/有机玻璃，也没有透明防紫外线膜，真空密封后的画作要远离窗子和日光灯以延长保色期。

If you do not have anti-UV glass/acrylic and do not have anti-UV film. Vacuum sealed picture should be displayed in a location away from window and/or fluorescent light to prolong its color retention.

装裱 Framing

材料	Materials
1. 镜框 2. ATG 胶带 3. 专业装裱镜框背后用纸（或牛皮纸） 4. 螺丝与挂钩 5. 镜框用铁丝	1. Frame 2. Hand apply ATG tape (Adhesive Transfer Tape) 3. Dust cover paper (or craft paper) 4. Frame screw and hanging set 5. Frame wire

真空密封与装裱 Vacuum Sealing And Framing

清洗玻璃或有机玻璃，放置在镜框内。如果你贴了防紫外线膜，膜的一边朝上（面对作品）。如果使用防紫外线玻璃，涂层向上。

Clean the glass or acrylic. Place the glass/acrylic in the frame. If you have applied anti-UV film on the glass/acrylic, the film side is facing up. If you are using anti-UV glass, the coating side faces up.

把纸框放入镜框，背面朝上。如果第一步的玻璃是普通玻璃，你也可以把纸框交由装裱公司覆膜而不用第一步的玻璃。

Put the mat on glass. Back side facing up.

把真空密封好的作品放入镜框，背朝上。扶好作品从正面检查看看是否一切都好，有问题马上修改。

Place the vacuum sealed picture in the frame. Check now to see how everything looks from the front. If you see anything wrong at this point, fix it now.

放一片薄海绵或者两张厨房用纸在画上面。

Put a thin piece of foam or 2 pieces of kitchen towels on top of the picture.

盖上背板。

Place the back board in the frame.

 用专业打钉机打装裱钉。
Use a framer's point driver to secure the points.

7 在木框四周贴上ATG胶带。
Apply ATG tape (adhesive transfer tape) on all 4 sides of the frame.

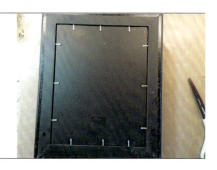

 撕下ATG胶带的保护纸。
Peel off the protection paper from ATG tape.

9 镜框后面盖上镜框后背纸或牛皮纸。
Cover the entire back with the dust cover paper.

用专业裁边刀裁掉多余的纸。如果你没有这个工具,也可以使用尺子与小刀。
Use an edge trimmer to trim off excess. You can also use a ruler and knife to do this if you do not have the professional trimmer.

真空密封与装裱 Vacuum Sealing And Framing

11

挂画的环需要钉在大约从上面算1/3长度，以红色箭头标明。用一个锥子先刺一个洞以方便拧螺丝。

The hanging rings are to be placed about 1/3 way down from the top where red arrows are pointing. Use a sharp tool to punch a hole for the screws to work.

12

用螺丝钉固定挂画的环。

Use screws to secure the hanging rings.

13

缠绕好铁丝。

Fasten the hanging wire.

14

在作品下面两个角贴上墙壁保护绒布，装裱完成。

Glue a wall protection pad on the bottom two corners. Framing is finished.

贴近看！美丽的东西可能很小。
——伊曼努尔·康德
Look closely. The beautiful may be small.
——Immanuel Kantm

07 压花设计

Pressed Flower Design

芍药近观
Peony Close-up

材料	Materials
1. 压花芍药连叶	1. Pressed peony with leaves
2. 12.7cm×17.8cm 水彩纸	2. 5×7 inch watercolor paper
3. 真空密封材料	3. Vacuum sealing materials
4. 镜框	4. Frame

鲜花组合 Pressed Flower Design

1. 把花瓣按照大小分类，剪一块大约5厘米×2.5厘米椭圆型的薄纸。

   Sort petals according to size. Cut an oval of thin paper about 2 × 1 inch.

2. 在椭圆纸上贴6片大花瓣。不要把花瓣贴得过分整齐，有一瓣有翻折会让花看起来更自然。

   Glue 6 large petals onto the oval paper. Do not arrange these petals super neat. Having a petal folding like this example would make the flower look more natural.

3. 再贴一圈花瓣。花瓣需要和下面的错开，花瓣顶端稍微往中心缩一点，让下面一层露出一点头。然后再贴半圈在花的上半圈。

   Glue one more layer all around. The petals are interweaved and the inner petal edges slightly inside to allow the 1st layer petals to show. After the 2nd layer, glue another layer on the upper half of the flower.

4. 粘贴花心。

   Glue the stamen.

压花艺术　Pressed Flower

贴一些小花瓣，让花心只露出头。修剪一下花瓣。大约需要3~4瓣。

Glue some small petals to just let the center to show partially. Trim the bottom petal off. We need about 3~4 petals.

5

注意，蓝色虚线那里，所有的花瓣需要很平滑地排好才看起来像自然的花瓣挺起来。这个地方是让花瓣看起来好像站立在那里，像这张新鲜花图上绿色虚线那里标示的样子。

有些芍药花瓣多，有些少。你可以按照你手上的花增加层次。

Notice the blue dotted the cutting line, all the petals should line up smoothly to look like natural petal bending. This portion is the view of the petals as shown on the petals of green dotted line in this fresh flower picture below.

Some peonies have more petals, some less. You can make your own layers accordingly.

Note

每朵花都是独一无二的，重要的是理解花的结构和组合技术。按照相片或画剪花瓣拼贴的芍药是无意义的。 什么使压花艺术成为艺术？是通过你仔细观察手中的那朵花，你赋予它生命。 不会与下一朵芍药相同， 它也不会是一个僵硬的工艺照片式拼贴。 它会在你的作品里独特地绽放。

Every flower is unique. The important thing is to understand the flower structure and the assembly technique. It is meaningless to cut each petal to make a photo realism collage of the peony. What makes pressed flower art art? It is by your careful observation of the individual flower, you give it life. It will not be the same as the next peony. It will not be stiff like a crafting photo. It will bloom in your artwork as itself.

玫瑰花束
Rose Bouquet

压制小型玫瑰　Pressing Small Roses

选择刚刚开的花。小型玫瑰或多头小玫瑰都是做这个作品好的选择。把花剖开一半，去除种子和海绵体。然后从中间去除一些花瓣，需要留3~4层花瓣。

Get flowers that are just starting to bloom. Small roses or spray roses are good selections for this project. Slice the flower in half, take out the seeds and the white sponge like material. Also take some flower petals out from the center. You need to leave about 3~4 layers of petals. Make sure the desiccant boards are activated before pressing.

选择搭配花　Select Companion Flowers

选择常见伴随玫瑰的小花。例如满天星、蕾丝花、洋甘菊、小翠菊和情人草都是很好的选择。

Select a small flower that usually go with rose bouquets. Baby's breath, Queen Anne's lace, small daisy, small aster, and statice are good choices.

选择背景　Select Background

可以选择任何材质。背景不要太花，不能和玫瑰抢风头。

You can select any background material but it cannot clash with the roses.

制作步骤　Procedure

首先确定花束如何在作品中摆放。可以斜放在对角线上，也可以竖立。这里我就用镊子确定花束的整体线条。

First we need to determine how the bouquet will be presented in the picture. It can be on a diagonal line or it can be straight. Here I have it on a diagonal just like how my tweezers are lined up.

鲜花组合 Pressed Flower Design

按照主线，在顶部排列一些线条，以确定花束的高和宽。在中间稍低的地方用些绿叶或蕨类，以确定茎将在哪里延伸。用三枝梗确定中线，以及手柄的宽度。

Follow the main line, we put some line elements on the top to determine the height and width of the bouquet. We also use some green leaves or fern below the middle to determine where the stems will be extended. Use one stem for the main line, two stems on each sides to determine how wide the stems will go.

3

开始粘贴玫瑰花蕾。
We can then start arranging the rosebuds.

花束需约12朵玫瑰，它们需要有一条想象的线可以延伸到下面手柄部分。任何不遵循假想线的花会看起来不自然。然后花束插入些情人草。

We need about 12 rosebuds for this bouquet. All of the rosebuds need to have an imaginary line that are extending to the handling stems below. Any flower not following the imaginary line would look odd since it is would not be physically possible. Insert some statice through out of the bouquet.

5

贴上一些梗。然后再贴好蕨，使从花到梗的那个过渡区看起来自然。写上落款日期完成。

We glue some stems. And then glue some fern to make the connection area of flowers and stems looking natural. Sign and date to finish.

康乃馨小花瓶静物
Still-life Carnation Vase

在这一部分，我们将学习如何压制和组合康乃馨，以及用硫酸纸制作磨砂效果花瓶的技巧。当然，也会进一步实践用软粉彩创作更多的背景色彩。

We will learn to press and assemble carnation flowers in this class. We will also learn some techniques on using translucent vellum paper to create a frosted glass vase. We will practice again using soft pastel for some background colors.

鲜花组合 Pressed Flower Design

压康乃馨　Pressing Carnations

1

切开花萼，剪下花瓣。浅绿色部分很难干燥，也没有什么作用，最好剪掉。可以用微波压花板先把花瓣压到九成干，然后再转用干燥板来压会加快压制速度。

Slice open the calyx, cut the petals out. The light green part of the petal dries very slowly and provides no real use. The light green part is cut off. You can try microwave the petals to 90% dried and then finish in the desiccant press to speed up.

2 把花萼和枝一起刨半，清理掉不需要的花瓣浅绿部分，刮掉枝中间的海绵体。每一半分开压。

Slice the calyx and the stem together in half. Clean out the unwanted light green part of the petals. Scratch off the spongy materials from the center of the stem. Press each half.

把花蕾剖半，从中间拔掉一些花瓣，留下边缘2~3片。
用干燥板压2~3天。

3

Slice the flower buds in half. Take out the petals from inside and leave 2-3 petals attached.
Press with desiccant boards. 2-3 days to dry.

组合康乃馨　Assemble Carnation

● 盛开花45°角 Full Bloom at 45° Angle

用3片花瓣形成一个扇型。用一丁点胶把它们固定。如同我们在初级书里学的，最好是用花胶或者中性硅树脂胶。

Take 3 petals to form a fan shape. Use a little bit of glue to glue them together. It is best to use either resin or silicone glue as we have learned in the first book.

加上第二排花瓣，注意顶部要比较接近。
再贴一个花瓣扇面。

Add a second layer to the fan shape. Notice that the top of the petals are close.
Make another petal fan shape.

把花瓣下面剪成弧形。
如果喜欢，可以加一层花瓣。注意花瓣顶要靠近。

Cut the base of the second petal fan in a curve.
If desired, we can add another layer of petals. Make sure the edges are close together.

把剪好的花瓣扇面放在之前的花瓣扇面上，如图。

Place the cut petal fan on top of the first petal fan as shown.

鲜花组合 Pressed Flower Design

两半连接处用花瓣填成自然过渡。
Place a petal on the side to make the two halves connect smoothly.

5

6 贴在花枝连花萼，完成。
Glue on a stem with calyx to finish.

● 盛开花正面　**Full Bloom Front View**

用纸剪一个比花瓣稍微大一点的圆形。
Cut a small paper circle just slightly larger than a petal.

1

2 在圆形纸上贴5片花瓣形成一个圆形。注意，这个圆形不能十分圆，看起来才比较自然。直径比两片花瓣稍微大一点。

Glue 5 petals on the paper to form a circle. Notice this circle is not perfect in order to look natural. The diameter of the circle should be slightly larger than 2 petals.

压花艺术 Pressed Flower

3

用4~5片花瓣形成一个内圈。注意，两圈的花瓣顶端比较靠近。
Glue 4–5 petals to form an inner circle. Notice the top edge of the petals are close together.

4

找3片内层花瓣，尾端剪成三角形，如图。
Find 3 inner petals and trim the bottom petals into triangular shape as shown.

5

把3片花瓣贴在花中间，尖角对着花中心。
Glue the three petals to the center of the flower with sharp angles pointing to the center.

6

用一两片花瓣调整花心，让它看起来自然。用小的有折的花瓣，把剪的痕迹隐藏起来。
Adjust the center with one or two small petals to make it look natural. Avoid exposing cut edges by using folded small petal to hide the sharp cuts.

鲜花组合 Pressed Flower Design

● 盛开花侧面　Full Bloom Side View

1

把几片花瓣贴成扇型。
Glue a few petals to form a fan shape.

2

盛开花的三个角度。
Altogether, full blooms in 3 different views.

● 花蕾　Flower buds

1

花蕾压出来是这样。
Flower buds are pressed like this.

2

把露在花萼外的花瓣剪掉即可得到花蕾的样子。
We need to trim the sides so no petal is seen outside of the calyx.

压花艺术 Pressed Flower

制作画 Making the Picture

材料	Materials
1. 水彩纸（20.3cm×25.5cm）	1. Watercolor paper 8×10 inches (20.3×25.5cm)
2. 软粉彩	2. Soft pastel
3. 白色油粉彩	3. White oil pastel
4. 粉红硫酸纸	4. Pink translucent vellum paper
5. 压花康乃馨，铁线蕨，满天星	5. Pressed carnation, maidenhair fern, baby's breath
6. 真空密封材料	6. Vacuum sealing materials
7. 装裱材料	7. Framing materials

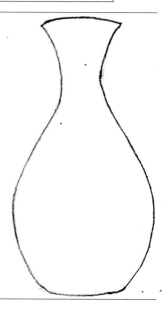

1

把这个形状描出来。

Trace this shape onto a piece of paper.

2

把硫酸纸放在形状的后面，然后剪下形状。

Put the vellum under the shape and cut the shape out.

鲜花组合 Pressed Flower Design

3

用白色油性粉彩画几条线,让花瓶看起来更有趣和立体。我们用反面,看上去更像毛玻璃狀。

Use white oil pastel to paint a few lines making the vase shape appear more interesting and look 3-D. We will use the back side which looks like frosted glass.

4

把软粉彩的粉末刮到水彩纸上,用棉花球均匀地把色彩涂抹开。

Scrape soft pastel powder all over the watercolor paper. Use a cotton ball to rub and distribute the color evenly.

5

用浅土黄画一些光线的线条,用手指顺着方向稍微晕开。

Paint some light earth yellow to suggest light. Use finger to smudge color according to the line direction.

在废纸上刮一些深粉红的粉末,注意不要在压花画上用荧光色。如果太粉太红,可以混一些浅土黄来调低一些。用棉花沾粉末然后涂出阴影。

Scrape some deep pink powder on the side. Do not use neon colored ones in pressed flower pictures. If it is too pink, mix some light earth yellow to tone it down. Dip a cotton ball in the pool of pink powder and then rub it on the background to form shadow.

6

压花艺术 Pressed Flower

7

首先放置花瓶和最高的花朵，确定画面的比例。接下来我们用花形成不等边三角形。正面的一朵是焦点。

First, position the vase and the tallest flower to determine the scale of the picture. Next we form a scalene triangle with the flowers. The front facing one is the focal point.

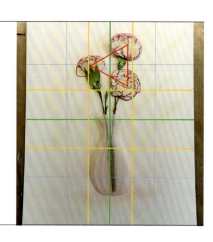

8

插入铁线蕨让花瓶里的花饱满。注意，这里我们还是运用不等边三角形的技巧。

Insert the maidenhair fern to make the arrangement full. Notice that it is also utilizing the scalene triangle technique.

9

插入一些满天星完成。

设计提示：枝条要从花瓶中伸展出来看起来才自然。

Insert some baby's breath to complete the arrangement.

Design Note: Make sure the stems are extended from the vase to look real.

Note

现在我们制作了两幅静物，可以看到：一个好的作品，需要各种不同角度的美丽压花。在压制之前，仔细地以不同角度观察鲜花很重要。

Now that we have made two still-life pictures, one might noticed that a good picture depends on beautiful pressed flowers with many different angles. It is important to observe the fresh flowers in many different angles before pressing.

向日葵
Sunflower

选择刚开的向日葵，其花心比较平，葵花籽还没有形成。

Select sunflowers that just started to bloom with the center relatively flat and without seeds.

压花艺术 Pressed Flower

压制向日葵　Pressing Sunflowers

● 正面花　Front View

1 压正面花，把后面的萼片摘除，保留最靠近花瓣的一层。这层不能有任何间断，不然花瓣会脱落。

To press front facing flowers, take out the green sepals in the back but leave one layer next to the petals completely intact. It is very important that you do not have any gap here or petals will fall off.

2 用小刀削掉后面一些皮肉，但不要过深，避免让花受损。

Use a knife to cut out some plant material in the back as shown. However, do not cut too deep and damage the flower.

● 侧面花　Side View

1 把花连梗一起剖半。

Slice the flower in half along the stem.

2 把花心切除，但保留靠花瓣部分大约1厘米。这能够防止花瓣脱落。

Take some of the center out but leave about 3/8 inch (1cm) toward the petals. This will prevent the petals from falling off.

鲜花组合 Pressed Flower Design

3

把梗的中间海绵体切除。
Cut out the meat of the stem.

Note

确保干燥板在压花之前还原好。 在向日葵花正面使用一层海绵，背后使用两层海绵。 用约4.5千克压力压制。 在半天后还原干燥板。 24小时后还原干燥板。 除去在花正面的海绵，增加压力到6.8千克。在第3天重新还原干燥板，将压力增加到9千克。 在第4天重新还原干燥板。从花后面取出一层海绵。 将压力提高至11千克。 留在压花板中一共6~7天。 花应在7天内干燥。 如果你的花很大，尝试更频繁地还原干燥板。压好的向日葵花心大约5毫米厚。

Make sure the desiccant boards are reactivated before pressing. Use one layer of foam for the front of the flower and two layers of foam for the back. Press with about 10 pounds weight (4.5 kg). Try to reactive the desiccant boards after 12 hours. At most, reactive the desiccant boards after 24 hours. Remove the foam in front of the flower, increase the pressing weight to 15 pounds (6.8 kg). Reactivate the desiccant boards at day 3 and increase pressure to 20 pounds (9 kg). Reactivate the desiccant boards at day 4. Remove one layer of the foam from the back of the flower. Increase pressure to 25 pounds (11 kg). Leave it in the press for total 6−7 days. Flower should be dried in 7 days. If your flower is very large, try to reactivate desiccant board more often. Press dried sunflower center is about 1/4 inch (5mm) thick.

制作方法　Procedure

材料	Materials
1. 压向日葵	1. Pressed sunflowers
2. 23cm×30cm 黑色厚纸	2. 9×12 inch black heavyweight paper
3. 镂空模板	3. Stencil template
4. 金属色粉	4. Metallic pigments
5. 粉彩定稿胶	5. Pastel Fixative
6. 真空密封材料	6. Vacuum sealing materials
7. 镜框	7. Frame

　　我扫了一些金属色粉在模板上面，在黑色背景纸上形成很漂亮的花纹。喷定稿胶，这样色粉就固定了。

压花艺术 Pressed Flower

I brushed a little bit metallic pigment powders onto different areas of the template. It forms a very attractive background on black paper. Spray fixative to make the patterns permanent.

贴花材的时候只用一点胶，能固定就可以了。要确保完成的画作放在有还原好干燥剂的密封盒子里，然后再真空密封。

Glue the flowers, stems, and leaves with just enough glue to hold the materials. Make sure to leave the finished work in a box with active desiccant overnight before vacuum sealing.

Note

你需要在作品后面放置两层棉垫来抽真空，以便多一些衬垫。
You will need 2 layers of cotton padding in the back of the picture for vacuum sealing for extra cushion.

鲜花组合 Pressed Flower Design

郁金香小花束
Small Bouquet of Tulips

鲜花组合 Pressed Flower Design

压制郁金香　Pressing Tulips

有两种压法。一是拆开花瓣，一瓣一瓣地压。另外就是剖半。

There are two ways to press the flower. One can press the petals individually. Or you can slice the tulip in half.

切除花蕊。用小刀刮梗子中间，把水分擦干。

Take out the stamens. Use a knife to scratch the inner stem and blot off moisture with a paper towel.

如果梗比较粗，去掉中间的肉，并擦干水分。

If the stem is thicker, take out some "meat" from the center. Blot off moisture.

把大叶子切一半。因为这些叶子弯曲不平，如果不切，会形成皱褶。

Cut the large leaves in half. These leaves are curvy. It will form folds if you don't cut it since the leaf is not flat.

压花艺术　Pressed Flower

5

用细砂纸轻轻擦叶子背面。

Use fine sand paper to sand the back side of the leaves lightly.

> **Note**
> 压之前一定要确保干燥板干透，第二天需要再次还原干燥板。花和叶应该三天压干。
>
> Make sure the desiccant boards are activated. Re-activate the desiccant boards on the 2nd day. Flowers and leaves should be ready in 3 days.

薄花处理　Extra Step for Thin Flowers

郁金香和很多球根类花一样，压后很薄。下面的步骤不是必须，但可以让花材的模样比较好，制作时拿取也比较方便。

Tulip and many bulb flowers are thin after being pressed. This extra step is not mandatory but will improve the flower handling and appearance.

1

将双面胶粘到和花瓣颜色匹配的薄纸上。或者你也可以使用白色的薄纸。普通纸太硬，使花瓣看起来不自然。

Adhere double sided adhesive to a sheet of tissue which matches the color of the petals. Or you can also use white tissue. Regular paper is too stiff, making the petals look unnatural.

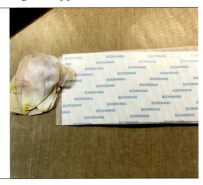

2

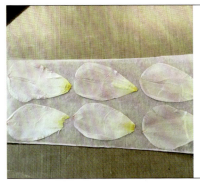

3瓣内侧朝上，3瓣内侧朝下放置在双面贴上。花瓣内侧色彩比较艳，不反光，外侧比较反光。

Place 3 petals inside facing up and 3 petals inside facing down on the double sided adhesive. The flower petal matte looking and with deeper color is the inside. Petal facing outside is shinier.

鲜花组合 Pressed Flower Design

3

把花瓣剪出来。衬了薄纸的花瓣更富有弹性和韧性。

Cut the petals out. The tissue backed petals are more flexible and stronger.

组合郁金香 Assemble Tulip

不管是拆瓣压还是切半压的郁金香都需要不同程度的组合。

Whether the tulip was pressed petal by petal or in half, it still needs some assembly.

● 有角度的侧面 Side View With Angle

1

上面的花瓣是内侧朝上，下面的花瓣是外侧朝上。把两片下面的花瓣如图剪半。

The top petals are inside petals facing up. The bottom ones are outside facing up. Cut the two side petals in half as shown.

2

把两片一半如图贴起来。另外两半也一样贴好。

Glue the halves together as shown. Repeat for the other halves.

3

把重叠的花瓣尖稍微修剪圆滑，让它们看起来更自然。

Cut the tip of the folding petals to slightly round the sharp line, making it look more natural.

4

上面的3片花瓣如图所示用少量胶粘好。

Arrange the top 3 petals as shown and apply some glue to secure.

5

把下面的花瓣放在上面，花瓣顶端比上面花瓣低一些,用一点胶粘好。

Arrange the lower petals on the top with the tips slightly lower than the top 3 petals as shown. Apply glue.

6

用剪刀把下端如图所示修圆。

Round the bottom with scissors as shown.

鲜花组合 Pressed Flower Design

7

梗贴在花后面完成。

Attach a stem on the back to finish.

● 侧面　Side View

1

所有的重叠花瓣需要小心地如图所示摘下来。

Any overlapping petal needs to be taken off from the stem carefully as show.

2

半朵外侧花瓣朝上，半朵内侧花瓣朝上，用浅粉红薄纸和双面贴如同之前一样贴在花的背面。然后把花形状剪出来。

Place the half flower petals, one inside facing up and one outside facing up, with double sided adhesive on light pink tissue paper just like for petals. Cut the shape out.

零散或剪半的花瓣都可用同样的方法处理。取半朵外侧花瓣朝上的，后面垫上内侧花瓣朝上的半朵，调整好花瓣，让后面的可以和前面的花瓣稍微交错。

Any loose petals, even those that were cut in half, are also treated the same way. Take an outside petal facing up backed with an inside petal facing up half to form a complete side view tulip. Adjust petal positions so the back petals are peeking out from the front.

小郁金香花束设计　　A Small Tulip Bouquet Design

郁金香的弯曲茎自然地融入了C形设计。 可以做一个C形或反转的C形。按照自然曲线，我们有一个非常可爱的浪漫花束。 为了凸显浅色郁金香，我采用了一个强烈的对比黑色的背景。

The curving stems of tulips naturally provides us a C shaped design. One can do a C or a reversed C. Following the natural curves, we have a very lovely romantic bouquet. To show off the light color of tulips, I have a black background to provide strong contrast.

在小型设计中保持线条清晰和简单。 不要过分挤入太多的郁金香。 你需要观众能够一眼先看主花。

Keep the lines clear and simple in a small scaled design. Do not overpower by crowding too many tulips. You need the viewer to be able to focus on the main flower.

搭配的花材，我们可以选择一个或两个有茎的。无茎的正面花朵不适合这种设计。配材我使用了同瓣草和小雏菊。 至于叶子，我使用了蹄盖蕨。

For companion flowers, we can select one or two with stems. Full faced flowers are not suitable for this design. I am using Isotoma and small daisies. For foliage, I am using lady fern.

先固定好主要线条，再添加搭配花材。选择花茎与主茎自然流向相同的。

Work on the lines first before adding more companion flowers. Choose the flowers with stems naturally flowing in the same direction.

蝶舞玫瑰园
Butterflies in Rose Garden

压花艺术 Pressed Flower

压花 Pressing

复习一下压小玫瑰。我们把梗剪下来，需要剪掉一点子房但不要剪得太秃，这样花才能保持整朵。

花蕾需要剖半，从中间取出子房和一些花瓣，留下3~4层花瓣，子房里面的白色物质去除。分别压。

绿叶和红叶都要压一些，也有的需要连枝压。

Let's review how to press mini roses. We cut the stem off at the base of the flower. Cut some of the bulging part off but not too close to the base so the flower will still hold together.

The rosebuds are sliced in half. Remove the seeds and some petals from inside. Leave about 3-4 layers of petals. Remove the white spongy material inside the seed pod. Press each half.

Press both green and young red leaves. Press some with stems.

背景 Backgound

这是一个两层的压花艺术设计。背纸是28厘米×25.6厘米的水彩纸（63.5千克）。上层是一张手染春雨落水纸（见背景一章）。

我们把春雨落水纸揉成一团，然后再展开，熨斗设置在化纤，把纸烫平。

这步是为了把纸里的硬纤维打软。

This piece has 2 layers of pressed flower art design. The backing paper is 11×14 inch (or 28×35.6cm) watercolor paper (140 lbs or 63.5kg). The upper layer is a piece of raindrop lace hand dyed (see chapter about backgrounds).

We shape the raindrop lace paper into a ball and then extend it out. Set a clothing iron to the polyester setting. Iron the lace paper flat.

This step is needed in order to soften the stiffness of the fiber in paper.

鲜花组合 Pressed Flower Design

设计要点　Design Notes

　　看起来自然，在花园设计中非常重要。 我们不是在谈类似照片的现实主义，但我们需要创造一个独特的自然的设计，以展示压花艺术的美丽。 因此，理解玫瑰在花园里如何生长是很重要的。 通常，一棵小玫瑰上有许多花同时开。 一株植物下部的叶子是成熟的绿色。 植物的尖端有新芽，是发红的。

Looking natural is very important in a garden design. We are not talking about photo realism but we need to create a unique natural looking design that showcases the beauty of pressed flower art. Therefore, understand how roses naturally grow in the garden is important. Usually, there are many flowers blooming at the same time on a given plant. The leaves on the lower plant is matured green in color. The tip of the plant has new growth which is reddish.

　　在造景设计中，我们通常应用三分之一法则。 这意味着在图中，我们有一个1/3的假想线。 上面的1/3为天空或下面的1/3是地面。 前人总结出，这会产生最具视觉吸引力的构图。

In landscape design, we usually apply the rule of one third. This means in a given picture, we have an imaginary line of 1/3. Either 1/3 makes the sky or 1/3 being the ground. This will yield the most visual appealing composition.

正如我们以前学到的，我们设计时必须首先考虑到主要焦点。您希望观众先看到什么？焦点是你的画要表达的。在这张画中，标题是蝶舞玫瑰园。我想向观众呈现的焦点是飞舞的蝴蝶，玫瑰花园是次要的位置。因此，蝴蝶在图片的中心。3只蝴蝶形成一个圆圈，使用类似的设计元素创造和谐和动感。玫瑰形成同心圆，进一步强调蝴蝶。

As we have learned before, we must always design with the main focus in mind. What do you want the viewer to see first? The focal point of the composition is always at where your theme is. In this picture, the title is butterflies in the rose garden. The focal point I want to present to the viewers is the dancing butterflies. The rose garden in this design is the location which is secondary. Therefore, the butterflies are in the center of the picture. 3 butterflies form a circle, using similar design elements creating harmony and movement. The roses form a concentric circle, further empathizing the butterflies.

玫瑰的安排并不都是随意的。在哪里放玫瑰是有目的地设计出来的。黄色最亮眼，它被放置在中间。它会吸引观众更多的注意中间。比较暗的在两边。

The arrangement of roses are not all random. There are some design thoughts in where to place roses. Of the colors in this arrangement, yellow draws the eyes first. It is placed in the middle. It will draw the attention of the viewers to the middle. The darker ones are on the sides.

双层设计步骤　　Procedures For Layered Work

在水彩纸的底部1/3处贴叶子。两边稍微翘一些。注意，虽然我们要形成一个弧形，我们应该遵循玫瑰在花园里的样子。它从来不是那么规则和局限。因此，我们需要有意地使叶安排看起来更像一个花园，松散地遵循圆弧形。

Glue some leaves on the bottom 1/3 of the watercolor paper. Raise the sides slightly. Notice that although we are going to form a circular curve, we should follow how roses naturally look in the garden. It is never regular and restricted. Therefore, we need to intentionally make the leaf arrangement looking more like a garden by loosely following the circular form.

鲜花组合 Pressed Flower Design

随意贴一些玫瑰。不是那么完美的玫瑰可以在这里使用，因为这一层的意图是创建深度效果。玫瑰花蕾应该安排在叶的上面边缘。

2

Glue some roses randomly. The not so perfect roses can be used here since the intention for this layer is creating a depth effect. The rosebuds should be more on the rim of the arrangement.

将春雨落水纸铺在设计的上面，然后排列花和叶。注意，不是所有的花都会面朝你。需要像在花园里一样，有些花会藏在叶子的后面。

3

Place the raindrop lace on top of the design, arrange flowers and leaves. Notice, not all flowers are supposed to face you, just like in a garden. Some are behind leaves.

Note: although this piece is intended for vacuum sealing, if you are not going to vacuum seal, it is best if you spray glue after step 2, and then glue the raindrop lace paper on the top before gluing pressed flowers for step 3.

● 正面蝴蝶　Front Facing Butterfly

用压花制作蝴蝶的方法很多。这只是其中一个。

There are many ways to make pressed flower butterflies. This is just one of the ways.

1

取一朵中型三色堇，把花瓣如图分解。

Take a medium size pansy and separate the petals as shown.

2

把下面的花瓣剪成两半。最上面的两瓣移到中间接长翅膀。

Cut the bottom petal in half and use the two from the top to extend the wings.

3

蝴蝶翅膀如图所示排列。然后我们需要制作触角和身体。

The butterfly wings are arranged as shown. Then we need to make the antennas and the body.

4

● 我们可以采取两个长而狭窄的嫩铁树叶或一些雄蕊，如牡丹的雄蕊等来制作触角。

● 一个真正的蝴蝶触角不会像很多卡通绘图那样卷曲。它与翅膀的比例也不是那么长。虽然我们不是做一个现实主义的图，但我们的意图是制作一个可信的压花设计。因此任何异想天开的设计都不合适。

● 用压花制作蝴蝶注意事项：蝴蝶翅膀是对称的。不要左右使用不同一朵花，因为没有两朵花是一模一样的。

Notice that a real butterfly antenna does not curl like in many cartoon drawings. It also is not that long in proportion to the wings. Although we are not making a realism picture but the intent is to make a believable one. Therefore any whimsical design is not suitable.

Note about making pressed flower butterflies: Butterfly wings are symmetrical. Do not use petals from different flowers. No two flowers are the same.

鲜花组合 Pressed Flower Design

● 侧面蝴蝶　Side Facing Butterfly

拿一个底部的花瓣，背面用顶部的深色花瓣衬好。如图红色圆圈所示，剪下来一个薄楔形。另外，剪一半的侧瓣为底翼。蝴蝶的肚子在侧面必须有一些曲线，如红色所示。

Take a bottom petal and back with the dark petal from the top. Cut a thin wedge off as shown in red circle. Also, cut half of a side petal for the bottom wing. The stomach of the butterfly on the side has some curve as indicated in red.

完成蝶舞图　Finish the Picture with Buterflies Dancing

蝴蝶应该是俏皮的，全部面对圆的中心。这样会把能量集中在中心，保持观众的注意力。

The butterflies should be playful and all facing the center of the circle. This is to focus energy toward the center keeping the viewers' attention there.

成功的压花艺术是展示天然材料的独特性，以充分发挥其潜力，并提供艺术家自己的信息和感觉。这不是仅仅为了好看。艺术家应该在制作一件艺术品时，时刻记得他的/她的情感，想法和信息是什么。

A successful pressed flower art is to showcase the uniqueness of natural materials to their fullest potential and also deliver the artist's own messages and feelings. It is not only about aesthetic beauty alone. The artist should always keep in mind what his/her emotions, ideas, and message are about when making a piece of art.

欢迎光临花园时光系列书店

中国林业出版社天猫旗舰店　　　花园时光微店

扫描二维码了解更多花园时光系列图书

购书电话：010-83143594